THE FOUR-DIMENSIONAL HUMAN

THE
FOUR-DIMENSIONAL
HUMAN

Ways of Being in the Digital World

Laurence Scott

W. W. Norton & Company
Independent Publishers Since 1923
New York · London

First published by The Random House Group Ltd.

For information about permission to reproduce selections from this book,
write to Permissions, W. W. Norton & Company, Inc.,
500 Fifth Avenue, New York, NY 10110

For information about special discounts for bulk purchases, please contact
W. W. Norton Special Sales at specialsales@wwnorton.com or 800-233-4830

Manufacturing by RR Donnelley, Harrisonburg
Production manager: Beth Steidle

ISBN: 978-0-393-35307-5 pbk.

W. W. Norton & Company, Inc.
500 Fifth Avenue, New York, N.Y. 10110
www.wwnorton.com

W. W. Norton & Company Ltd.
Castle House, 75/76 Wells Street, London W1T 3QT

1 2 3 4 5 6 7 8 9 0

To Stella and David Scott,
with unending love and gratitude.

All the birds are singing,
and the day is just beginning,
Pat feels he's a really happy man.

'Postman Pat and his Black and White Cat'

CONTENTS

INTRODUCTION

The Reverse Peephole

———————

I was glad that the visit to the mock-castle had been a success, since I'm a mock-tourist. My friend and I were sitting on the terrace of our bed-and-breakfast near Sintra, Portugal. It was the spring of 2008, which means that there were perhaps Couchsurfers down in the valley, but Airbnb was several months shy of its founding. We had just been on a trip to Quinta da Regaleira, the imitation palace, which in the early twentieth century a set designer had built for a coffee and gem merchant. Although I had enjoyed crawling through the watery tunnels on the castle's estate, I was happier still to be back on the terrace, writing a postcard.

The card was a thank-you note to an acquaintance who, a month or so previously, had lent me his empty house. He hadn't known I was coming, since he was already abroad when I emailed to ask him. His Georgian dwelling on a lonely terrace was not without its Gothic qualities. Every other chandelier worked, through the barred windows in the dining room lay a forgotten garden, and all the internal doors opened except one, on the second-floor landing. As I turned and shook the handle, it rattled in its frame. I wondered what he kept locked in there, even when no guests were expected. Elsewhere there were touching signs of the absence into which I'd intruded: a bowl and spoon on the desk, smeared with dried yoghurt, maybe, or ice cream, a magazine left on the floor by the bath, folded to an article titled 'The Loveliest Doors'. During my stay I had the recurring dream that he came back suddenly to turn me out, full of a mysterious and inarguable anger.

Nevertheless, thanks were due. I finished the postcard and set it aside. My friend, the good traveller, put down her book and we started chatting. As we spoke, I noticed that I kept glancing at the postcard. Its presence on the table was a sort of agitation, gathering around it a subdued but recognisable cluster of feelings. There was a general nervousness and sense of waiting. For a few moments on the warm terrace, with the forested valley below, life was somehow incomplete, as though a bite had been taken from it. Towards my hospitable acquaintance with the broken chandeliers I felt both impatient and afraid of having offended him, a blend that settled itself into a mild strain of resentment. The card was belly-up on the table, drying its ink in the sun. All this wordless agitation, taking me away from my friend and our jokes, voiced itself suddenly in the thought: 'Why hasn't he answered it yet?'

In retrospect it seemed as though, in that deranged moment, I had wakened to a process that had been quietly rewiring my life for a decade, more or less since I chose my first, cryptic email address (imagine broadcasting my real name on 'the internet').

The fiction of this period reflects a wider fascination with the idea of other dimensions reachable from our own – worlds that appear as enigmatic glimmers, or else are unreliably accessible, through doorways that don't always open, or which refuse to stay fixed in one place. In Joseph Conrad and Ford Madox Ford's co-authored novel *The Inheritors* (1901), a man called Granger meets a visitor from the Fourth Dimension. The novel begins in Canterbury, where Granger and the Dimensionist are crossing through 'the old gateway'. They gaze down at the Gothic city, when suddenly the visitor utters a strange sound and the buildings below them start to change: 'One seemed to see something beyond, something vaster – vaster than the cathedrals, vaster than the conception of the gods to whom cathedrals were raised. The tower reeled out of the perpendicular. One saw beyond it, not roofs, or smoke, or hills, but an unrealised, unrealisable infinity of space.'

A hundred years after its heyday, we're only now inhabiting space in a way that could be called four-dimensional. When the early, domestic internet appeared in the 1990s, there was a decisive separation between physical reality and that other place, which went by several aliases. The first household modems enforced this separation by acting as though they were grinding up against something hard, squealing and whirring like a drill hitting rock. Sitting next to them, perhaps wearing a t-shirt over a long-sleeved t-shirt, we might have sighed or yawned, looking mildly around the room or prodding at the squidge of the mousepad, but all the while feeling quickened somehow, as on the verge of arrival. The modem's faithful churn made it seem as if it were tunnelling through to somewhere else, opening up a space for us to inhabit. Once inside we followed our moods, web pages listlessly completing themselves in descending strips, producing all manner of suspense as the news story or piece of erotica toppled slowly into being. And then someone in the house would need the telephone, the ultimate, old-world trump card. A voice would rise up the stairs, and the tunnel would cave in, Pete Sampras's Australian Open fate hanging in the balance.

So the modems gave the sense of a journey. Through certain designated portals we could move into a specific way of being that felt like entering a new territory. By gaining access to this landscape on the other side of things, our bold *fin de siècle* achieved on its own terms a feat of which its predecessor dreamed. But the scene from *The Inheritors* suggests how discovering the fourth dimension isn't an experience that you can segregate from everyday life. The tower reeled, the mundane spaces warp and twist. The fourth dimension doesn't sit neatly above or on the other side of things. It isn't an attic extension. Rather, it contorts the old dimensions. And so it is with digitisation, which is no longer a space in and out of which we clamber, via the phone lines. The old world itself has taken on, in its essence, a four-dimensionality. Every moment, every object, has been imbued with the capacity for this extra aspect. Just as a geometrical net of squares can be folded into a cube, our daily lives are a series of nets, any of which could be scored and bent at the perpendicular, and thus extended into this other dimension. Increasingly, the moments of our lives audition for digitisation. A view from the window, a meeting with friends, a thought, an instance of leisure or exasperation – they are all candidates, contestants even, for a dimensional upgrade.

Social media, for example, makes a moment four-dimensional by scaffolding it with simultaneity, such that it exists in multiple places at once. A truth and cliché of digital life is that our comeliest meals occur both on our table and in the pockets and on the desks of our international 4D colleagues, a meal to be both eaten and approved of. One quality of the fourth dimension is co-presence, life happening both locally and in the mind of someone elsewhere. With the prospect of this digital fourth dimension, a moment can feel strangely flat if it exists solely in itself. Then again, to think of a moment as being 'flat' is really a throwback from the three-dimensional days. Today, we live with the sense that un-tweeted, un-instagrammed moments might feel somehow cubic, as in *boxed in*, just these four walls, unless the walls can be contorted along invisible lines and a message smuggled out. Few people have trouble finding such a smuggler now; it's a mass industry, this smuggling of life into four dimensions.

Digitisation's warping and renting of old-fashioned space has inevitably affected our notions of domesticity, the architecture of our privacy. With walls not being what they once were, the home itself has become four-dimensional, with new ground plans to match its digital environment. Increasingly we are said to 'live' online, to create virtual habitations and communities for ourselves, but our physical homes have also been digitised. We can identify a common fitting on a 4D house by travelling back in time to the unlikely world of *Seinfeld*'s last season. In an episode that first aired in 1998, Kramer and Newman, two characters not known for their level-headedness, decide to reverse the peepholes in their apartment doors so as 'to prevent an ambush'. The idea is that, on returning home, they can check if an assailant is waiting inside to 'clock them with a sock full of pennies'. Seinfeld himself is incredulous when he hears about the reverse peephole, pointing out that 'then anyone can just look in and see you', to which Kramer replies, 'Our policy is we're comfortable with our bodies. If somebody wants to help themselves to an eyeful, well, we say, "Enjoy the show!"'

What strikes me most about this scene now isn't the craziness of the plan but the innocence of Seinfeld's objections. He can't conceive of anyone wanting to fold their homes inside out in this way. His ideas of doors and walls and windows, those ancient technologies of seclusion and revelation, through which we have historically negotiated the basic bounds of publicity and privacy, are thoroughly three-dimensional. The reverse peephole is, in this sense, visionary in its anticipation of the digital revolution, a definitive anxiety of which is that our peepholes have been reversed without our knowing. But an arguably more interesting phenomenon is the voluntary reversal, for now we endorse and facilitate all sorts of peepholes into our domestic interiors. It is perhaps during our drowsy meanders of the deep night, alone in the glowing dark, that we most often find ourselves, through social media's chain of associations, in a kitchen full of strangers, caught in a moment of togetherness. One could rightly argue that these views are stage-managed, a show to be enjoyed, the opposite of an ambush. And yet there's always an excess that can't be controlled, knowledge that slips around the sides of the spotlight. This is a new vision of our homes, with windows opening onto faraway rooms, and lights shining out into remote darknesses.

The Four-Dimensional Human

A dominant idea in the history of modern western personhood has been that we're fundamentally isolated from one another. Certain key figures, real and imagined, are gathered together, no doubt reluctantly, to make this argument: Descartes sitting in solitude at his fireside, sceptical of everything but his own mind; the fantasist Don Quixote, alone in his suspicion of windmills; the Princess of Cleves, moving between the seclusion of the boudoir and the convent; Robinson Crusoe on his island. That being an individual entails a sort of exile from others may be a story that we tell ourselves, but it is no less solid for that. Of course the irony here is that we also can't seem to get enough of the pack. We gather our lonesome selves together in groups by day, clinging together in warm, mealy huddles by night. Yet no matter how tight the clinch, we're still flung to different corners of the dreamscape.

In *A Tale of Two Cities*, Dickens writes of 'A wonderful fact to reflect upon, that every human creature is constituted to be that profound secret and mystery to every other.' An unnamed 'I' ponders the inscrutability of those closest to it, realising that 'No more can I turn the leaves of this dear book that I loved, and vainly hope in time to read it all.' The image of a person as an unreadable book also appears in Edgar Allan Poe's story 'The Man of the Crowd', in which one man follows another through the streets of London. This tale's horror lies in the follower's conclusion that he can never know the stranger he is following: 'It will be in vain to follow, for I shall learn no more of him, nor of his deeds.' He imagines the stranger's heart as being larger than a medieval book of prayer, and thinks how it ultimately may be 'but one of the great mercies of God' that it 'does not permit itself to be read'.

If the ultimate solitariness of the modern person has been a well-told story, then how does this story contend with the spirit of the digital twenty-first century? As I consider this question, I pause too often to do a lap of that standard-issue circuit of websites, turning the keys in those letterboxes and jogging past the lives of others, who stand smiling and telling me things as I go by. The year is ending, and people on Facebook are posting a prefab review of the past twelve months. 'It's been a great year!' the official script says. 'Thanks for being a part of it.' I click on one of these retrospectives, posted by a friend on whom neither the corniness nor the irresistibility of this venture is lost. This friend, as represented in her string of aggregated pictures and captions, is a collage of other people, a luxuriating pet, interspersed between picturesque landscapes. Social media is an advocate of this brand of comradeliness, and encourages us to narrate our lives as legibly as possible, as ongoing books that invite themselves to be read.

At the same time, the various and non-stop opportunities for communication are notable for highlighting our isolation, and it's perhaps this intensity of digital communicability that brings mythic proportions to mind. When the Olympian postman Hermes goes in search of Odysseus during the latter's long confinement on Calypso's island, he looks for him in a cave, but 'Of Odysseus there was no sign, since he sat wretched as ever on the shore, troubling his heart with tears and sighs and grief. There he could gaze out over the rolling waves, with streaming eyes.' I think of this weeping Odysseus sometimes, when I'm waiting with indecorous zeal for an email or a text, or when I catch myself peering into the rolling blue of Facebook, unable to remember for whom or what I'm looking. I see his yearning in miniature, in the five seconds it takes for someone to bring a phone from their pocket and put it back again. These are ship-in-a-bottle feelings, which life can accommodate. The otherwise cheerful and productive of us have cheerful, productive lives amid digital longings and desolations. But it is certainly true that invoking the messenger god is one of the constitutive practices of our times. It has become part of the rhythms of almost every waking hour, to look for a word or a sign from elsewhere. We want to feel the wind in our faces, with the full, oxygenating sense that we're coursing along. Going online can feel like a step on a homeward journey, where it is the abstract promise of home, rather than any real sense of the home itself, which matters. We all know the pocket-sized shipwreck that occurs when an inbox shows us, with treacherous indifference, the pale, empty horizon of read emails.

The shipwreck-in-a-bottle is one of many new digital phenomena that have become part of our daily experience, and which complicate age-old ideas of personhood. It has long been the word on the street that, if you dabble in other realities, then you shouldn't expect to remain unchanged. Lazarus was never his old self again. Visiting fairyland has its chronic side effects and implications: unnatural youth, blindness, contraindicated with trips through the wardrobe. Our portals to the fourth dimension have been wedged open, and there it is, spread out across the everyday, indeed nestled inside the everyday, causing it to ripple and bend. And now that the silhouette of a figure is resolving in the doorway, fringed in ghost-light, we can begin to consider what might be on their minds, what it feels like to be flushed with the hormones of Web 2.0. What new senses are available to someone who is such a concentrated blend of matter and media? What happens to the nervous system when it is exposed to the delights and pressures and weird sorrows of networked life? How does time pass in this dimension? What dreams begin to prey on a four-dimensional mind? What are the paradoxes and ironies of owning a four-dimensional body, with its marvellous new musculature?

A crucial tension of our times is that, although we can luxuriate in this gained dimension, stretching our lives into the world like never before, we are simultaneously asked to ignore, deny, accept, strategise or rail against the hypothesis that our physical planet is diminishing. Just as the fourth dimension is opening up before us, our old-world trio is, by all intuitive accounts, in crisis. The macabre package of images and arguments, grouped under the deceptively benign term 'climate change', ticks like a telltale heart beneath the fibre-optic cables. A discernible atmosphere has emerged in these times from the collision between the digital boom and the ecological bust. One prevailing style of online being suggests a tireless lust for life. Social media steers us into co-producing a catalogue of daily rapture. The tacit caption beneath uploaded photographs is, not infrequently, 'Behold!' There's a general intoxication over the well-framed moment, and we have a hunger for beautiful vistas that was not a pre-digital appetite. People younger than me look up at a building hit with late-afternoon light and think, 'That's 'grammable.' We say a new sort of grace with the click of a camera and give thanks for our loved ones, a scrum of more or less willing smiles. We share good songs and good writing; we like all manner of propositions and support one another. Yet in between this digital fervour, we have little choice but to live with the apocalyptic sentiment in our water supply.

Looking back, those few weeks of house-sitting in 2008, with their memorable air of consent and trespass, now seem like a 3D rehearsal of a commonplace 4D pastime. Every morning I would walk by the locked door on my way to the office. The study where I worked could only be accessed through a tight passageway that connected it to the office. The study had bookshelves on three walls and a large window on the fourth; the urban bleakness of the view was ideal. Between bouts of writing I prowled the small room. My presence there was both invited and unexpected, and in such circumstances every movement hovers between snooping and entitlement. I made a rule for myself very quickly that I would read nothing on the cluttered desk that wasn't already visible, and that I would stop reading the moment it would be despicable not to. Inevitably, hidden things rose to the surface. Cards fell from books; books had personal inscriptions. I remember a wedding invitation being used as a place-mark, which described each course of the meal in detail. My absent landlord emailed me from abroad to ask me to look in his desk drawer for some paperwork. The drawer will have car keys in it, he said, and some private letters. Please do not read them. The certainty that nothing would compel me to open those letters made me warm with a sudden sense of my own goodness.

I didn't think too much about that locked room, since its mystery became entwined with my general unease in this grand, eerie house. The study's strange layout made it a sort of snare, and if I worked in there at night I waited for the sound of soft footsteps at the passage's mouth. In the mornings I was hard on myself, but not too hard. It was in these evening hours that the idea of the locked room would be most present, and occasionally I would sit in this trap of books and lamplight and wonder what inhabited that space, just across the way. A vague notion of 'electrical goods' came to me, an unlikely bank of computers. I pictured exercise equipment, a Stairmaster, sets of dumb-bells and a mirrored wall, locked away as a rule, being too incongruous even for their owner to stumble upon freely. The prospect of a gimp's studio of course crossed my mind (live and let live), but mainly the contents of the room existed as an amorphous blank, a gap in my knowledge of things. Then one morning I was at the desk in the study and something inchoate made me turn in my chair and run my hand across the wall of books behind me. I turned back to see the tame, daytime guise of that narrow little passageway, so inconvenient and improvised. My mind retraced the journey back through the passage, into the office and out into the landing, where the locked door protected its secrets. I smiled at my idiocy, for it was suddenly clear that, all along, it had been me inside the forbidden room. The desk's disarray, my papers and his papers, the computer and the rows of books, even the white sky outside, became vivid with realisation. I had been chasing myself. During those evening hours I had been simultaneously inside and outside that room. In an instant, all of my theories about its contents fell in on themselves, and the blankness was imprinted with a sudden picture, a selfie before its time.

there were jeroboams in corners, little secretive lamps and vases of full-faced flowers, all doubled in the congested reflection of a large gilt mirror. The vistas were as thickly detailed as an illuminated page from the Book of Kells, which lay on its back nearby. White was reserved for the tablecloths alone, spaces cut out for drinkers to fill with their own crowded talk.

We pushed two central tables together and our tutor ordered champagne. You would have to have been a fool not to have thought highly of yourself. The sense of having arrived somewhere was strong among the youngest of us. La Cave, with all its stuff, was undeniably *there*. But then my thigh fluttered. Guiltily I dilated my pocket under the table and peered inside. The first stranger to whom I'd ever given my mobile number, a South African Spar manager, was prowling about in the city above us and thinking of me. Another dimension of my new Irish life was cleaving the overdressed walls and perfect camaraderie of our evening. The champagne arrived and was poured, but my mind was up and gone in the dark streets, not simply dreaming but fixing plans. Even the most fledgling desire can wing its way out of heaven. It's hard to believe how, at that time and in certain situations, a mobile phone could still be a curio, and certainly the practice of sustained, barefaced texting was as much of a novelty in millennial Dublin as a South African managing a Spar. At the roof of my vision I saw our generous tutor turn to touch his glass with mine and, after clocking the sordid finger-work going on in my lap, turn away again. Few social moments corral people as effectively as a cheers, and equally few are as fracturing as a spoiled one. One of my classmates laughed at what I was doing, and over his shoulder my tutor said paternally: 'Don't be a cunt, Laurence.' This scolding, like many fatherly injunctions, made its way into my superego, such that, for a long time afterwards, being caught texting in polite company had a genital impropriety to it.

Being called a cunt in La Cave was the primal trauma of my four-dimensional infancy. Those first years of the new millennium could be remembered as a period when we began to see rents in our habits, to feel tears in the fabric of everyday life that revealed a new sort of world behind them. As the walls of La Cave cracked and cleaved around my questing mobile, our small party was being exposed to and remarking on a normalcy of the near-future. My puncturing of our sociability, the bleeding away of presence, was judged an injury to the ambience. For they'd caught me shimmering between two places, living simultaneously in the 3D world of tablecloths and elbows, and also in another dimension, a lively, unrealisable kind of nowhere, which couldn't adequately be thought about in the regular terms of width, depth and breadth.

Those of you beginning to settle into this book's dimensional conceit shouldn't be overly worried that the 1959 sci-fi movie *4D Man* was renamed *The Evil Force* for British audiences and *Master of Terror* in America. The film's mad scientist learns to pass his hand through a block of steel and is eventually able to walk through walls. He runs amok with his newfound skill, but the evil force propelling him isn't the ability to reach into the bank's coffers. Although the power is unhealthy, accelerating his aging and generally wearing him out, the source of his wickedness is less his four-dimensionality and more the timeless tragedy of his 'girl' falling for his brother. What seems superficially to be a morality fable about corrupting power is really a two-dimensional love triangle. For our purposes this is just as well, since we might think of ourselves in a digital age as possessing something comparable to the 4D Man's ability to slip through solid objects.

The 4D Man's four-dimensionality is a corporeal event, which results in a power as simple as opening a door. His altered body, indifferent to physical boundaries, becomes like a skeleton key capable of unlocking matter itself, such that he is no longer restricted by three-dimensional laws of confinement. Digitisation may have prompted an analogous event in us, but our story is happily more complicated than a B-movie. If our bodies have traditionally provided the basic outline of our presence in the world, then we can't enter a networked environment, in which we present ourselves in multiple places at once, without rethinking the scope and limits of embodiment. While we sit next to one person, smiling through a screen at someone else, our thoughts, our visions, our offhand and heartfelt declarations materialise in fragments in one another's pockets. It's astonishing to think how in the last twenty years the limits and coherence of our bodies have been so radically redefined. We have an *everywhereness* to us now that inevitably alters our relationship to those stalwart human aspects of self-containment, remoteness and isolation. Like the 4D Man, we are able to insubstantiate ourselves to the point that the solid stuff around us seems insubstantial. Unlike this jealous genius, however, we can be on both sides of the wall at once.

We often trade notes on our engagements with digital technologies as though discussing remarkable physical feats. Recently I was giving a lunchtime talk about the undeniable weirdness of life in the digital age, when a stately gentleman set aside his lasagne to rebut the hint of scepticism that he'd detected in my voice. 'On the other hand,' he said, 'I have *face time* every week with my sixteen-month-old granddaughter while she's eating her dinner, and when she says goodbye she hugs the iPad.' Face time is apparently the new quality time, now that our faces have gone global. It seemed that, for the grandfather, this ability to participate affectionately in his granddaughter's evening routine was justification enough for any vertiginous feelings it might otherwise cause. Togetherness, no matter the format, is what counts. Several of his colleagues, many of whom were of his generation, accepted the benefits of family relations *sans frontières*. But, despite this endorsement, the weirdness angle of my lecture only gained momentum with the story of an infant in Connecticut hugging Pennsylvania Grandpa cum iPad. It's strange to have your heart warmed and chilled in one go. I was especially interested in how the child demonstrated love for Pop's familiar face and voice by clasping his proxy self: the lukewarm breast of a tablet. While the baby may not yet be up to making such distinctions, the question still remains whether the iPad is a part of Grandpa's body, an extreme extremity, or Grandpa himself, a rectangular, coherent presence hovering by the high chair.

Where do our bodies begin and end in a networked world? What sort of physique do we need to travel through two sorts of spaces at once, through the quaint streets, with their pavements, cars, overhead weather, brick walls and lamp posts slick with rain, and that other place, that strange land of everywhere and elsewhere, where our connectivity occurs? And, crucially, what does it feel like, for us and for others, as we move around inside these digitised skins?

As I burrow back into the spatial certainty of my remembered childhood, looking for answers to and helpful metaphors for this new situation, the figure of Mr Tickle suddenly appears. I was always spooked by Mr Tickle, particularly one illustration of his extensive, baked-bean-coloured arm snaking flaccidly through a bedroom window. 'Perhaps that extraordinary long arm of his is already creeping up to the door of this room,' Roger Hargreaves suggests, unhelpfully for any child at odds with the night. But now our arsenal of mobile devices makes Mr Tickles out of all of us, with the added upgrade of being able to push our tickling fingers, as the 4D Man did before us, through solid walls. Solidity itself, as we've discovered, is not what it used to be. The invisible channels of Wi-Fi and cellular networks have carved into the material world, and now the old enclosures are riddled with two-way escape routes out of the moment. We stretch fluidly down these fissures, emerging in other places, and, equally, far-off people extend themselves into instant proximity. And when we don't touch each other in real time, we're equipped with more or less friendly arrows and the gift of perfect aim. 4D air quivers with arrow-fire; we move through the world being skewered in the hip or the lap or the palm. Is this then one of our major new manifestations: a cross between Mr Tickle and St Sebastian, shot through with sociability?

The terrifying thing about the 4D Man is that nothing can restrain his body. This excess of freedom defines him against everyone else, and in our own case our 4D corporeality comes into focus when we think about how less containable digitisation has made us. In the recent past, much of social life *was* a form of containment: for better or worse, being stuck with others. The genre of jokes involving people walking into bars came before our digital everywhereness, and relies on the idea of a bar as a relatively closed system in which the patrons have no choice but to interact with one another. For those seeking a dalliance, bars were once improvisational spaces where you make the best of what you've got or, more glumly, get what you're given. Every one of us who has ever been a-huntin' in a nightclub knows when the force field separating two new lovers from the rest of the pack has been activated. If we've had any vested interest in either party, we can feel it powering up in the pleasantries mouthed into hair-covered ears, or in the particularly sharp jaw shadow of someone hornily straining to listen. But now with mobile apps like Grindr and Tinder, which provide an instant Rolodex of nearby hook-ups who are likewise alert to one's whereabouts, the force field may be seen to flicker and the territory can never be fully secured. The bar is, for one night only, no longer the world, and at any moment someone of more obvious interest may wander into one's wider cyber-radius. Messages from hotter prospects two streets away could suddenly fly in like drones, and the less lucky of the duo may be gazumped at the last.

It's clear how everywhere we've become when it is necessary, for sporting purposes, to rein us in. I was struck by the change to our inherent self-containment on hearing of the demise of the phone-a-friend lifeline on *Who Wants to be a Millionaire?* In 2010 the US version removed this option because, over the years, these friends had become little more than mediums for Google. As the presenter Meredith Vieira explained, almost everyone's friend was 'Mr Internet'. The UK franchise held on longer. In later series, when a struggling British contestant sent out the SOS, the host Chris Tarrant had to assure viewers that this Samaritan was in a specially isolated environment, removed from smartphone and internet. This proviso would have seemed bizarre when the programme began in the mid-1990s, and shows the extent to which humans in their homes no longer possess their once innate separation from the world and its oceans of facts.

A connected but stranger quizzing case, however, is that of Richard Osman's fake laptop on the afternoon trivia programme *Pointless*. As co-host, Osman sits behind a computer, providing various statistics and amusing facts after the answers are revealed. I once attended a taping of *Pointless*. Someone in the audience asked the warm-up comedian why the laptop was just a prop. 'Well otherwise,' the comedian shrugged, 'he'd just be some guy at a desk.' Osman has said that it 'gives me an air of authority'. As we filed slowly off the set and into the midsummer evening, I felt a pathetic undertow in passing Osman's empty seat and his dummy computer. Perhaps this feeling was linked to a subconscious pun, but the comedian's words evoked the Emerald City edict of 'pay no attention to the man behind the curtain'. In *The Wizard of Oz*, the Nebraskan charlatan is trying to hide his gadgets of deception in order to appear as an immense presence before his subjects. But *Pointless*'s minor act of fakery intends the reverse. 'Pay attention to the gadgets!' it cries. The aim is to simulate a person in our own modern image, not a pulsating, all-knowing head but rather someone visibly extending into those mundane technologies of omniscience. The fake laptop, in other words, completes a credible, fully realised silhouette of the everywhere-person.

The various video-chat platforms, of which Skype is an early tone-setter, enable the most vivid aspect of our everywhereness. Like the smitten Pennsylvania grandpa, millions of us daily take advantage of such conviviality, delighted to carry the severed heads of family members under our arms as we move from the deck to the cool of inside, or steering them around our new homes, bobbing them like babies on a seasickening tour. Skype can be a wonderful consolation prize in the ongoing tournament of globalisation, though typically the first place it transports us is to ourselves. How often are the initial seconds of a video call's take-off occupied by two wary, diagonal glances, with a quick muss or flick of the hair, or a more generous tilt of the screen in respect to the chin? Please attend to your own mask first. Yet, despite the obvious cheer of seeing a faraway face, lonesomeness surely persists in the impossibility of eye contact. You can offer up your eyes to the other person, but your own view will be of the webcam's unwarm aperture.

Everywhereness is not an easy business, for it can make life feel both oppressively crowded and, when its promise is wasted, uniquely solitary. If you work on a computer and are at the mercy of Skype's default settings – the devil's in the defaults – it's likely that your day is punctuated by people you know staking their flags into the moment. Sometimes, if you work at home, the timing may be unfortunate: your mother's standard, for instance, unfurling just as *Ji₂₂ Brat₂* is picking up steam. In calmer times, these short-lived alerts can be melancholy depending on one's mood, like watching the sails of far-off boats catching and losing the sun. At other times they're inconsequential, but isn't there something melancholy in feeling the inconsequentiality of other people's presence? The contact list of any messaging or video-calling system, a column of lit and extinguished lamps, inevitably dulls us to the miracle of connectedness. And isn't it strange to think of all those friends and loved ones poised behind a screen, like a stalled episode of *This Is Your Life*? So often, tact or shyness or busyness prevents us from making anything of this pseudo-proximity. Almost every day I watch the name of a certain close friend, with whom I've fallen out of touch, rise to obscure a corner of my toolbar, and each time I watch this little tag dissolve and wonder what she might be doing.

The novelist A. S. Byatt, born in 1936, has remarked that although she finds it odd to watch people in the streets absorbed in their phones, she concedes that overall they seem happier than strangers did in her earlier years. You only have to read Eugene McCabe's short story 'Music at Annahullion', which describes the crushing isolation of remote rural lives, to be eternally thankful for the succours of Skype. The unremitting, inescapable loneliness of three middle-aged siblings living in a desolate farmhouse in Ireland leads the sister to declare one interminable day: 'I wish to God we were never born.' One feels that a Wi-Fi router would have gone a long way at Annahullion. However, there's a tiring aspect to Skype that, given such morbid pronouncements, one feels churlish to admit.

The causes of this fatigue are difficult to identify, but for me the problem lies in the fact that we can't bring our silence with us through walls. In phone conversations, while silence can be both awkward and intimate, there is no doubt that each of you inhabits the same darkness, breathing the same dead air. Perversely, a phone silence is a thick rope tying two speakers together in the private void of their suspended conversation. This binding may be unpleasant and to be avoided, but it isn't as estranging as its visual counterpart. When talk runs to ground on Skype, and if the purpose of the call is to chat, I can quickly sense that my silence isn't their silence. For some reason silence can't cross the membrane of the computer screen as it can uncoil down phone lines. While we may be lulled into thinking that a Skype call, being visual, is more akin to a hang-out than a phone conversation, it is in many ways more demanding than its aural predecessor. Not until Skype has it become clear how much companionable quiet has depended on co-inhabiting an atmosphere, with the simple act of sharing the particulars of place – the objects in the room, the light through the window – offering a lovely alternative to talk. No doubt people have found ways of using Skype, not so much as the arena for chat, but more as a magic periscope, such that you can come and go, read and cook, or fold up laundry, while checking in now and again with your miniaturised friend. But the strange sorrow that Skype provokes has to do with the disorientating nature of being in multiple places at once. For where exactly do these encounters occur? In what here or what there do we meet? In your space or mine? An unspoken and alienating question in this not-quite-face-to-face medium can easily be: 'Is that really you?'

Creatures of Elsewhere

In the olden days, a person's capacity for everywhereness went by the name of imagination. Unlike our Mr Tickle ubiquity, the three-dimensional journey to everywhere tended to arise from solitude rather than from connectivity. Towards the end of the nineteenth century and into the next, canonical English-language writers became especially interested in dramatising what they saw as the essential solitariness of human nature. Of all the scenes in Henry James's *The Portrait of a Lady*, James was most satisfied with his description of Isabel Archer sitting up alone one night in her drawing room, the fire dwindling, thinking about the disaster of her marriage to Gilbert Osmond. What can we make of this favouritism? Technically this chapter was progressive because it contains almost no external action, propelled instead by the movements of Isabel's mind. Like a wildlife film-maker, James was pleased to have captured a fundamental but elusive aspect of a species' behaviour: the solitary human, self-absorbed, tunnelling inwards, unearthing unpleasantness and sorrow. In this scene Isabel communes solely with her own grim sense of reality. Her thoughts about her husband – intense ideas about his motivations and desires – produce a thick and oily portrait of him. 'He was', she decides, 'like a sceptical voyager strolling on the beach while he waited for the tide, looking seaward yet not putting to sea.' James dramatises how thinking about someone else's life is often a way of thinking about your own. While supposedly dwelling on her husband, this meditation makes Isabel remember that she was 'after all herself'. James's achievement here is to show the very human magic trick of being at once contained and roaming. While Isabel's body sits isolated in the drawing room, the candles burning low in their sockets, her mind is undeniably elsewhere. James tells us how 'Isabel wandered among these ugly possibilities until she had completely lost her way . . . Then she broke out of the labyrinth, rubbing her eyes'. This elsewhere leads, ironically, not to escape but to another form of confinement. She considers how Gilbert has trapped her in the mansion of his domination. 'Between these four walls', James writes, 'she had lived ever since; they were to surround her for the rest of her life. It was the house of darkness.'

In one sense, Isabel was just some gal by the fire. And yet she conjures up a rich, virtual world in the solitude of her drawing room, a world made possible by the absence of external stimuli. The presence of other people has traditionally barred us from the deep labyrinths of the mind. At the same time, her husband Gilbert is very much with her during those hours of contemplation, and his oppressive influence on her thoughts suggests how fragments of our existences live waywardly in the minds of other people. After James, modernist writers became hooked on this puzzle of the human as being, simultaneously, an introspective, secretive, enclosed consciousness and a communal project, something erected collectively like the marquee at a summer fete, or the Amish barn in *Witness*. Marcel Proust had clearly not conceived of anything remotely like Facebook when he wrote that

> even in the most insignificant details of our daily life, none of us can be said to constitute a material whole, which is identical for everyone, and need only be turned up like a page in an account-book or the record of a will; our social personality is a creation of the thoughts of other people.

Virginia Woolf's Mrs Dalloway, in the stretch of a morning stroll to buy flowers, thinks of herself as 'being out, out, far out to sea and alone' and also 'part of people she had never met; being laid out like a mist between the people she knew best, who lifted her on their branches as she had seen the trees lift the mist, but it spread ever so far, her life, herself'.

Despite recognising human interconnectedness, Woolf's characters are sensitive to its boundaries. 'What solitary icebergs we are!' Clarissa Dalloway's husband declares in his debut appearance in Woolf's first novel, *The Voyage Out*. In *Orlando*, the Russian woman Sasha is described as a cloaked entity: 'For in all she said, however open she seemed and voluptuous, there was something hidden; in all she did, however daring, there was something concealed.' Sasha's mystery refers in part to the inscrutability of her foreignness, but Woolf's overall notion of human relations seems to figure us as foreigners to one another, and beautifully so.

From her drawing-room window, Clarissa Dalloway watches her elderly neighbour moving behind the windows of her own house and thinks of the 'supreme mystery' of human solitude. She formulates this mystery, which she calls the 'privacy of the soul', as: 'here was one room; there another. Did religion solve that, or love?'

Can Skype solve that? It attempts to override this separation by placing one room inside another. Certainly, a popular rallying cry of social media is: you in one room; me in the same room. As Woolf would say, our lives spread ever so far now. Some would argue that all this extensiveness isn't *us* at all, and that we're still as confined as ever we were to the long-and-shorts of our bodies. For all the followers and the likes, the linking-ins and the retweets, the four-square points and viral felines, are we ultimately still solitary icebergs? In emphasising the revolutionary impact of the First World War on human experience, Walter Benjamin wrote how 'A generation that had gone to school on a horse-drawn streetcar now stood under the open sky in a countryside in which nothing remained unchanged but the clouds, and beneath these clouds, in a field of force of destructive torrents and explosions, was the tiny, fragile, human body.' But the cyber-revolution exploding beneath and within the Cloud, which is both ever-changing and permanent, seems to be achieving the opposite to tininess, fragility and solitariness. The children of digitisation will grow up expecting to occupy space robustly and to live prolifically in one another's rooms. Their strength will be measured, like the density of muscle fibres, according to the knit of their connectedness.

This big bold future, however, will demand an evolution in how we think about what it means to be present, how we manifest bodily and virtually in the world. When the writer and director Jonathan Miller was asked about his opinion on the afterlife, with signature bemusement he replied with a question that is an ongoing problem for the architects of our cyber-economy: 'How would it know it was me?' He continues:

> Everything about myself that I know is to do with the fact that I am embodied. My body makes me here rather than there . . . The fact that there is a here rather than a there, everything is to do with the fact that the hereness and thereness of things is determined by [my] body.

Miller's account of himself is a wholly three-dimensional one, a horse-drawn carriage of embodiment. But we are now witnessing a remapping of 'here' and 'there', in accordance with the go-go-gadget elasticity of our online selves. The pressures of everywhere-ness, which call for a collapse of here and there, can produce a sense of absenteeism, and the suspicion that, despite being in many places at once, we're not fully inhabiting any of them. This spatial etiquette is still under negotiation, and one illusion that we're continually perfecting is not simply how to be here and there at the same time, but how to be *there* while looking like we're *here*.

Consider this anecdotal evidence taken from my old-fashioned pastime of gawping at the people opposite me on public transport. In a Piccadilly Line carriage a mother is leaning sideways to argue quietly with her late-teenage son. I hear the word 'immature' in her whisperings. 'Mum, I'm not discussing this on a train . . . It's the same every time.' She mentions money; he seeks assurance from his free newspaper, from the sportsmen in his lap. Facing a wall of adolescent sullenness, she settles back in her seat. Presently, she begins to smile and leans in again. 'To change the subject completely . . . this is just silly,' she takes a piece of paper out of her bag and unravels it like a shopping list, holding it up in front of her son's face and reading him excerpts from it. I guess that he isn't the type to laugh at his mother's silly aside simply to smooth things over. His winter hat has red stitching in the shape of a tiny Che Guevara and he wears the adolescent's dog collar of puffy black headphones. The mother keeps reading aloud and chuckling, and he stares at the strip of paper as though watching from afar a mildly revolting but ultimately tedious spectacle. The mother's smile subsides. 'Not interested, never mind,' she says quietly, pressing her scarf against her breast, at which, perhaps, more gracious children once suckled. 'Paper,' she sighs, and begins to flick pointedly through the *Metro*, her eyes moving quickly over sections of each page beneath arched, overacting eyebrows. He stares desolately at his own copy. I can hear the white noise inside both of their heads as they continue this pretence at reading. Despite her bluff, the mother soon abandons the headlines and brings out her smartphone. She starts tapping at it thoughtfully while the son remains very still and holds his downward stare, waiting. It's not long before unselfconscious quivers of concentration appear at the edges of her mouth. Now and then her eyebrows tremor from true absorption and focus, and a new sense of peace washes over the scene. The son tugs his jeans at the knee and adjusts his weight; he begins to turn the pages of his newspaper naturally. I wonder where this sudden ease has come from, and then I realise that it is because the mother has gone.

To absorb fully your attention in smartphone perusing, as any astrophysicist of the psyche will tell you, requires a much lower escape velocity than does reading, which in comparison is a rocket-straining endeavour. The mother beaming herself up into her inbox is a textbook side effect of our ability to be everywhere, namely that we may often seem elsewhere to those physically nearest us. It has become a part of the everyday rhythm of social life that we dart in and out of each other's view – here one minute, gone the next: 'I'm just checking . . .'; 'I'll just answer this . . .'; 'Bill Murray in *Ocean's Eleven*? Let's just see what IMDB has to say . . .' How long do those apologetic 'justs' have left? This intermittent elsewhereness has come to seem an ordinary aspect of human behaviour.

As Henry James tried to convey with his fireside scene, we have always been creatures of elsewhere, never confined to our own skins. Consciousness, you might say, is a passport, but traditionally we've gone travelling in our own time. Adolescents, notoriously, are frequent-flyers, with portable, personal music-makers being their standard teleportation devices. Although never a music buff – I don't think I knew a song before Brandy and Monica's 'The Boy is Mine' – in my more winsome years I would bop down the street with my earbuds in, and I was almost always transported into the most elaborate and self-serving of parallel universes. I would materialise in Wimbledon, inhabiting the body and conscious-ness of Martina Navratilova as she/I volleyed our way to fictitious victories. Or else the teleporter flung me inside the opening credits of a television show, where over and over I would spin around on a stool, my name appearing glamorously under my smouldering, famous face. Humbler reveries took place in nightclubs populated by everyone I had ever admired sexually, gathered to watch me dance. In by far the most shameless example, a catastrophe had befallen the city of Detroit, and as one of its major home-grown recording artists I was asked to take part in a benefit concert star-ring myself and two other native luminaries of a similar calibre: Madonna and Eminem. The three of us put on one hell of a show, and depending on my mood we sometimes kissed ostentatiously after each doing a verse of Eminem's 'Lose Yourself' in our own celebrated styles. In other words, I spent years poised at the entrance to elsewhere. Even the grimy twang of *Melrose Place*'s theme song could accidentally set me off on one.

As years passed my teleportation device became less reliable, and for a while I knocked in vain at the back of that particular wardrobe. Perhaps then it was with curmudgeonly envy that I baulked at a recent trend among young pro tennis players to arrive on court wearing their MP3 players. Victoria Azarenka, Andy Murray and Sabine Lisicki are among those on my headmaster's list of misbehavers. The environmentalist is trained to chart colossal change in the small things: in the lowered decibels of frog-song, in a derelict swallow's nest, in a blizzard of monarchs reduced to papery flurries. As a flat-footed also-ran of the Ontario junior tennis circuit and nine-time imaginary Wimbledon champion, I'm apt to notice wider changes to our self-containment in the hermetic environment of a singles tennis match. Indeed, the sport of tennis generally, being dominated by notions of solitary glory, is a good place to look for evolving expectations of presence and absence, and the insistency of everywhereness.

Historically, tennis has been an exhibition of intense self-containment, in which players are concentrated beings of the here and now. The theatre of a singles match – one of the most solitary events in sport – demands that its actors be alone but not absent. In the arena, a singles player's solitude is paradoxically a communal phenomenon, a part of the shared drama, and in days of yore both players and spectators were quick to feel any interference with this essential isolation. During the 1999 French Open final, when Martina Hingis crossed the net to question a ball mark on Steffi

Graf's side of the clay, the crowd jeered at Hingis's transgression of her symbolic separation from other people, and the umpire gave an extra point to Graf as a penalty. At the same tournament ten years previously, Monica Seles offered a bunch of flowers to her opponent Zina Garrison at the start of the match. The two commentators for Eurosport felt that Garrison would be right in lodging an official complaint. Full catharsis at the end of a match depends on isolation immediately before and during it; hence the tragic power of Hingis being half carried back onto the court by her mother after her eventual loss to Graf, or Jana Novotna's notorious moistening of the Duchess of Kent's shoulder.

The wearing of headphones to the court tampers with this spectacle of isolation. By the end of 2013, the top two female players – Serena Williams and Victoria Azarenka – had sponsorship deals with Beats by Dre headphones. Williams, I notice, keeps hers around her neck during her entrances, but then she hails from an earlier time. A personal music player is not purely a digital technology, but its arrival in the tennis stadium suggests digital life's effect on our notions of presence. Although there were Walkmans and Discmans in the 1980s and 1990s, I don't recall players ever bringing them onto court. It's possible that they do so now because, in the 4D world, presence is often partial. It doesn't seem as radical for modern players to spirit sections of their attentions away, given that in everyday life we are decreasingly materialised, less singularly attentive. There was indeed anxiety when the Walkman arrived, which had to do with the potential blurring of public and private lives. The idea of people cocooning themselves inside songs while walking the streets or riding the bus took some getting used to, and was viewed as being predominantly the vice of escape-obsessed teenagers. This practice is commonplace now among many age groups, and we have since moved on from the music player's clear declaration of antisociality to the momentary cocoons offered by the digital age's mobile distractions. We may have always been daydreamers, but in the past if we lapsed into a daydream while in company, people would perhaps take it as a sign either of gaucheness or of madness, and we may have had a hand waved in front of our faces. Today we allow each other to travel back and forth from elsewhere within the stretch of a conversation, moving in and out of our physical bodies before each other's eyes.

Body Incorporated

Everywhereness was always an agenda of the early pioneers and prophets of the web, though they imagined online life not as being one of widespread, simultaneous embodiment, of Grandpa in and not in the kitchen, but rather the opposite: unincorporated experience, or life without a body. By going online, people used to think, we were shrugging off our physical selves in exchange for an immaterial existence. Indeed, the prospect of turning into a bodiless being was one of the initial excitements of online life. In 1996, the writer and activist John Perry Barlow voiced one school of optimism for what the internet might achieve in 'A Declaration of the Independence of Cyberspace'. This open letter was addressed to 'Governments of the Industrial World', warning them that their traditional ways of disciplining and controlling a populace would have no purchase in the immaterial, online sphere. 'Your legal concepts of property, expression, identity, movement, and context hat into the ring. The Microsoft slogan was more existential than perhaps it intended, since an original promise and allure of the internet was the space it offered for self-invention and indeed self-inconsistency. People could surf along on undulating versions of themselves, 'Where Do You Want to Go Today?' becoming 'Where Do You Want to Vanish?' Alternatively, a shadowy self in the real world could be for the first time illuminated by the online affirmations of kindred spirits. In other words, dual or even multiple lives were free to proliferate. Of course the corollary of all this existential liberty wasn't so much the crisis of 'Who am I?' but rather the fear of 'Who are you?' An instant archetype emerged of the old pervert plaiting his cyber-pigtails and skipping into schoolgirl forums. Those who were so inclined could come online with an army of what became known as sock puppets: fake profiles run typically from a single source, designed to give the illusion of support in any chat-room skirmish. I myself dabbled briefly in these heady freedoms, sharing a chair with my friend Clara in her dorm room and playing an online, acronym-based parlour game under the joint name of Sugar. I remember two things from these misadventures: acronyms referencing the Clinton–Lewinsky affair always scored well, and banter with fellow players tended to deteriorate swiftly ('Go to hell, Sugar').

These online experiments with identity matched the wider mood of the 1990s, a decade interested in debunking supposedly rigid identity categories, particularly those concerning gender and sexuality. Alongside 'On the internet, nobody knows you're a dog', there began to be a mainstream enthusiasm for the idea that identity was too fluid for discrete labels such as masculine and feminine, gay and straight, dog and not-dog. The problem, however, became one of political visibility: how do gay people, for example, through our collective actions.' In late 2013, the UK government's National Cyber Security Programme unrolled its Cyber Streetwise campaign. Cartoonish adverts on Tube carriages show people leaving foolish instructions for would-be thieves: a note on a door saying 'Keys Under Pot', or lipstick scrawled on a window: 'Valuables in Bottom Drawer'. The official website's animated short videos suggest the absurdity of our online behaviours when transferred into 3D scenarios. In one of them, a woman leaves the photo-printing shop with a set of snaps, and then distributes them around the clientele of a nearby cafe. The website explains how the campaign 'underlines that safety precautions taken in the real world have similar relevance in the virtual world'. This scheme is on one hand an aspect of the government's civic responsibility, and there can be something almost consoling in a touch of mild paternalism. However, an implication of its chiding slogan 'You wouldn't do this on the street. Why do it online?' is the conflation of reality and virtuality. Did we imagine, during the first whisperings about this exotic thing called the internet, which in the mid-1990s began to possess certain chosen desktops in the school's computer lab, that it would ultimately create an analogue to life, a collapsed replica of the world that we inflate with each journey inside and which contains many of the same concerns and dangers?

An arguably more relentless and successful lobby for online embodiment is commercially, rather than civically, motivated. If we return to how online life has been sold to us over the years, it becomes clear that on or around 2011, web browsers changed their pitch. In 2011, Google Chrome introduced its challenge of a tag line: 'the web is what you make of it'. In one ad called 'Jess Time' a motherless young woman (#heartstrings) survives the first turbulent months of dorm life, every major detail of which she seems to recount to her father via Google video. The ad's jubilation lies in the fact that Jess has both travelled afar and gone nowhere, like the shuddering surrealist image of a train that hurtles out of Gare de Lyon and yet never leaves. In another, a lovelorn man sends an emotionally manipulative multimedia package to his ex-girlfriend that is meant to persuade her to rekindle their affair. He includes links to pictures of old date-spots, video footage of them tipping over the crest of a roller coaster, even a Google-map of the bench where they broke up. Here Chrome depicts itself as a rehabilitator and redeemer of the past, its role being not to transport you into the unknown but to archive the treasures of personal history. Another subgenre of this campaign has parents assembling online baby books of their children's early years. This browser isn't there to take you somewhere; it is to remind you who you are. These documented children have a YouTube presence before they can read.

Google's fantasies present a conservative view of self-formation couched in family and old bonds, while more radical transformation is reserved for commercial adventures, such as in the advert 'Julie Deane', where the real-life heroine starts an online satchel business and does very well. Likewise, another ad depicts a young couple surviving the wrath of Fannie Mae and Freddie Mac to establish their own hot-dog restaurant. Yet even in showing the web's entrepreneurial possibilities, Chrome is eager to imbue these successes with a comforting, bathetic homeliness. Julie Deane's satchels have branches in London, New York and 'Our Kitchen', while the budding sausage moguls are referred to as 'Mom and Pop'. Here you get the reach of the global brand mingled with the comforting smallness of private life.

It has behoved the merchant class to promote this embodying and containing of our cyber-selves. The ideal consumer is, after all, not a shape-shifter but a stolid type with consistent and predictable habits. In other words, we're asked to be as reliable as the things we buy and in this sense there's little difference between a consumer and a commodity. Both must stand still and lock eyes in order for the match to be made between desirer and desired. Hence that moment in so many movie shopping montages in which the shopper stands transfixed by something on the other side of the window. If we turn into a fluctuating, amorphous species online, then how will people know what products they should aim at us?

Consumerism has traditionally thrived in the real world from our having fixed, demographically analysable identities. For this reason its agents are understandably keen to minimise the differences between our material and digital lives, to erode the boundaries between the real and the virtual. It's business-as-usual in the fourth dimension. Take, as just one example, the 2014 online campaign by skincare company Nivea, in which a guys' guy darts around his apartment demonstrating tricks for the on-the-go dude: putting smelly gym shoes in the freezer, or an ice cube in the clothes dryer with dress shirts to help with creases. All this is dispensed with locker-room breeziness, butching up the atmosphere in readiness for the final piece of advice: smear Nivea skin cream on your face like a granny with her Oil of Olay. Tellingly, the ad refers to these tips by the now-common term 'life hacks', as though it makes sense to think of your morning routine as a sort of computer program, with efficiency and organisation being equated to a hijacking of the old code. 'Wanna make the most out of life? Then hack it', Nivea urges.

If a cyber-street is more or less equivalent to a real street, and our gym shoes can be hacked, then in this merger the bones and blood of us will also have their counterparts online. We might be able now to be everywhere, and therefore are forgiven for sometimes being elsewhere, but our digital selves are not the mercurial creatures that many believed they were destined to be. The drive for our online embodiment comes from a civic and commercial conservatism, which, far from the radical possibilities of the early internet, seeks to reproduce real-world notions of personhood, of the stable, predictable, measureable consumer-citizen. When this conservatism fuses with the true radicalism of digital life, namely the multiplicity of our presence in the world, one outcome is that we begin to manifest according to the logic of the brand. Online we can stand on many platforms at once, but we are encouraged to be the same wherever we go, our personal logo stamped across the net. Instead of being the shape-shifting surfers of an independent cyberspace, it is expected that we be franchises, that we make chain stores of ourselves.

Airbnb and the Chain-Store Self

One of my memorable Facebook status updates went something like: 'How long before newsreaders realise that we all know that the social networking website Twitter is a social networking website?' I got a lot of good feedback from that one, including someone I used to fancy prodding his thumb up at it. The day took on a subtle shine, and I kept returning to Facebook to check how my little aperçu was doing, like looking through the oven window at a rising cake. The news may now be on first-name terms with Twitter but, as I write, the same cannot yet be said for the private holiday rental platform Airbnb. When telling vacation stories, people will still say to me, 'Do you know that thing . . . Airbnb? We used that.' However, Airbnb is poised for common parlance; it will no doubt soon lose the red, corrugated training wheels of Microsoft's Spell Check.

Airbnb is one of the increasing number of cyber-services that demands its users have fixed, 4D bodies – a chain-store identity. It is an apt illustration of how we are asked to materialise online as contained, knowable people, to enter into a community of fully realised digital subjects. We are required to accumulate an online history of consistent, amiable personhood, so that we can be recognised wherever we crop up in digital space. This paradigm rings a death knell for the bodiless shape-shifter of the early web, despite one of Airbnb's slogans echoing the spirit of the first popular web browsers: 'The only question is: where to go next?' A common reason that people travel is to try to outrun their old lives, to experience fresh starts, to be blissfully anonymous, a mysterious stranger blowing through town. But in order to journey with Airbnb, we are required always to bring ourselves with us. Our pasts become a travel document; we're not allowed to shed our digital skins.

By the time these pages reach your eyes, providing a working definition of Airbnb may well be as useful as explaining what a vacuum cleaner does, but in case of some amnesia plague or an impending and terrible turn in the company's fortunes, I will say that Airbnb is a service that coordinates the temporary letting of people's homes to travellers desiring 'unique' and generally cheaper accommodation. The second 'b' of 'bnb' (I'm reminded of Homer Simpson's BBBQ) is largely ornamental, since the host is not contractually obliged to offer up a tempting array of fruits, cereals and yoghurts, with perhaps a cooked or continental option as well. So if you choose to enlist your home on Airbnb, yours will join the gaggle of rooms competing for attention in your neighbourhood. You must provide pictures of your interiors, as well lit and handsome as possible, much like the best-face-forward approach to online dating.

There's some degree of pathos in Airbnb's turning of the home inside out for financial gain and pitting it against the other items in the windows, which perhaps more reflects the western housing crisis than it does a utopian adoption of Bedouin hospitality. Hosts in New York City, one place where Airbnb has been most interrogated municipally, have claimed that the Airbnb supplement makes their rents affordable. Likewise, visitors say that they wouldn't have chosen to visit New York without access to such economical and pleasant accommodation. In some ways the concept of Airbnb is a practical, exciting evolution of the hospitality industry, but, to make such a business possible, what old instincts about the sovereignty of the home and about the stranger at the door have been outmanoeuvred? If you go back to around 2011, it's clear that these old instincts were still in place, since online discussions of Airbnb frequently drifted into talk of axe murderers. The format can produce a mutual paranoia in which both host and guest become both predator and prey. Fairy tales exploit the primordial vulnerability of occupying an unknown house ('the first bowl was too hot, the second too cold . . .'), as well as the reluctance to open up your own to strangers, particularly those bragging about their lung capacity.

Until recently I lived in the top flat two floors above a small-scale pizza takeaway, which made for impolitic moments when my Papa John's delivery pulled up outside. Between us there was a small dwelling that, when I moved in, was without a doorknob and padlocked shut. The occupants were a mother and grown-up son whose pastimes included chain-smoking and quarrelling vigorously. They liked me because in my craven way I agreed to them storing a bicycle that occluded eighty per cent of our bottleneck hallway. I saw them rarely and heard them always, a discordant symphony of footsteps, wailing and slammed front doors. On Valentine's Day they moved out, headed for a bigger place further west. 'It'll be nice and quiet now,' the mother said when I met her and her possessions on the street. 'This was only ever a temporary thing.' A van was parked by the pizza scooters with its back open, blinking its hazards. Two years later, letters still arrived for them, often of an urgent nature or from the NHS. The flat stood empty for months, and then a doorknob appeared and footsteps returned to the stairwell. On two occasions I bumped into two different people who gave me the same name and who turned out to be squatters. After the lesser-spotted cockney owner evicted them, months of vacancy would be interrupted by the comings and goings of unseen and furtive new inhabitants. On one of those unholy nights when you know that the sleep demon is standing at the foot of your bed, I holstered a rolling pin in one of my mattress's side-handles, entertaining wild thoughts about the latest mystery guest below.

During my second summer there, envelopes addressed to a professor began to appear. One morning I heard the yaps of a dog in the hallway. A writer had bought the flat of shadows, and she greeted me in a boiler suit, announcing that she was going to finish painting the floor and then go on *Open Book*. I tentatively returned the rolling pin to its drawer. Unsurprisingly the flat was only meant as a pied-à-terre, and during her absent times, the writer said, she would be accepting guests from Airbnb ('Do you know it?'). And so it was that my life continued to be perched above a thoroughfare, with none of the reassuring routines of constant neighbours. At startling times the main door would open and strange, woollen voices would come through the walls, except that now they were often joined with the huffs and strains of suitcase haulage. Indeed, while the revolving door continued to spin, a significant change had occurred. The place had been transformed from a secretive refuge for the temporarily dispossessed and the desperate – a black-hole asylum whose boiler never ran and where letters could be lost and demands evaded – into a spruce and transparent little crash-pad for the global traveller.

I looked up the flat's listing on Airbnb, and soon found the first name and picture of my professor, smiling politely in her good pearls. I could read both the reviews of her visitor-customers and her thoughts on them: the acts of kindness and mutual goodwill, as well as vivid details about one disastrous stay involving (separately) improper linen usage and a toilet bowl full of steeped urine. A man I had seen leaning against the tree outside the pizza shop was now in the gallery of client-friends, and by clicking on him I could see where he lives, goes to university, and peruse samples of his prose.

A key writer in the late-Victorian vogue for the fourth dimension was Edwin A. Abbott, whose 1884 novel *Flatland* is both a social satire and an allegory of inter-dimensional travel. In the story a three-dimensional being visits the two-dimensional plane-world of the book's title, which is inhabited by sentient lines and triangles and priestly circles. The 3D visitor, a sphere, describes to a Flatlander his higher perception of two-dimensional space: 'From that position of advantage I discerned all that you speak of as *solid* (by which you mean "enclosed on four sides"), your houses, your churches, your very chests and safes, yes even your insides and stomachs, all lying open and exposed to my view.' Airbnb relies precisely on this kind of exposure, a 4D scrutinising of our three-dimensional world. The flat below me had become like Flatland. Its ceiling had been blown away and I could, if I liked, peer inside it, see its tables and chairs and carpet without ever passing through the front door. And in the case of bad reviews, it is often the private messiness of the body that is revealed, its unsporting excretions and stains, the clots of hairs in the plughole that soil the reputations of slovenly guests.

The growing popularity of Airbnb testifies to our sense of every-whereness, which enables a feeling of continual connection to the safe and the familiar. Wherever we go, part of us is always at home. A thousand miles from our loved ones, we can pull a stranger's blanket up to our chin and manage not to feel eerie, soothed no doubt by the night lights of our phones and laptops. In this sense, our tentacu-lar digital bodies help us to defeat an age-old dread of being cut off from the familiar and cast into an unknown environment. These primordial fears have been famously depicted in the psychodramas of children's stories. The initial horror of the fairy tale 'Beauty and the Beast' is that it's a collision of two strangers under one roof. Having grown too tired to complete his long journey home, Beauty's father checks himself into the Beast's imposing lodgings, which are at once homely and unhomely, both comfortable and desolate. We are told how 'The pleasant warmth of the air revived him, and he felt very hungry; but there seemed to be nobody in all this vast and splendid palace whom he could ask to give him something to eat.' He eventually finds a room that is, in accordance with the magic of the place, vacantly hospitable, with an inviting plate of supper quietly steaming on its tray. In the morning, having still seen no one, he reasonably assumes that this disembodied generosity extends to his plucking a rose for his daughter from the palace gardens. It is at this point that he encounters the Beast, who is enraged by his thievery.

Since he is a bewitched prince, the Beast symbolises the concealed body. His true self is obscured by the spell, while, to him, Beauty's father appears to be a sticky-fingered trespasser. After hearing some desperate explanations and apologies, the Beast is merciful. 'You seem to be an honest man,' he growls, 'so I will trust you to go home. I give you a month to see if either of your daughters will come back with you and stay here, to let you go free.' Like many online entrepreneurs from the distant future, the Beast offers a month's free trial. The catch, however, is far worse than landing a second month's subscription at full-whack price. If the father can't convince a daughter to become the Beast's next and final guest, then he will have to surrender himself for ever.

This exchange is a prototype of what Airbnb's CEO Brian Chesky calls 'the trust economy', the growing range of digitally coordinated enterprises based on members of the public sharing space or cars or other typically private resources. Beauty's father is an anonymous guest and so the Beast's trust is a product of instinct and concealed gentility. However, as with the irony of today's commercialised version of trust, it is hollowed by the fact that the Beast's powers of surveillance and capture make trustworthiness irrelevant. The father's oath to honour this deal and the Beast's sensing of this honour are mere adornments to the transaction because of the Beast's final warning: 'Do not imagine that you can hide from me, for if you fail to keep your word I will come and fetch you!' The Beast is a creature of his fairy-tale world not only in appearance but also in his professed ability to short-circuit the anonymity of strangers. If you can be traced then you aren't anonymous. In a 2013 interview, Chesky describes the milieu in which Airbnb operates, while presenting the same contradiction as the Beast: 'You can call it the sharing economy. Or the trust economy. I think there's something really special about that. A year from now everybody [on Airbnb] will be required to verify, meaning share

33

their email and their online and offline identity.' Airbnb's million dollars of insurance coverage for each of its hosts, and the demand for user transparency, seem to indicate the opposite of trust, which by its nature is the sum of our reckonings with the unknown.

Disney's account of the fairy tale foresaw the perfect Airbnb stay in the song 'Be Our Guest', in which the Beauty, Belle, is serenaded by the anthropomorphised objects in the Beast's palace. Jigging crocks are a Disney shorthand for *joie de vivre*, and these crocks are delighted to treat Belle to a spectacular feast. 'No one's gloomy or complaining while the flatware's entertaining,' sings the candlestick. The motherly teapot can't wait to start bubbling and the champagne bottles are popping their corks, all of them revelling in a quasi-erotic desire to be used. Those Airbnb-ers who stay in the lived-in homes of intermittently present or fully absent owners experience a muted version of this hospitality-by-proxy. In such a model, the guest communes with household objects that act as avatars, the feathered plumpness of the duvet, the pedigree of the kettle, the heft of the cutlery all speaking to the host's virtue. In their Hospitality Standards section of the website, Airbnb tells us that 'Every day, hosts around the world create magical experiences for thousands of guests.'

Freud naturally had much to say about the home and the idea of homeliness. The German word for homely is *heimlich*, and Freud, with the help of others such as the philosopher Friedrich Schelling, emphasised a glitch in its various definitions. *Heimlich* means homely in the English sense of something or somewhere friendly, welcoming and convivial, emanating warmth and safety. However, the same word can also evoke notions of concealment, secrecy and even conspiracy. An obsolete German term for a privy councillor – a keeper of secrets – is *Der heimliche Rat*. This darker shade of home-liness arises from an image of the home as a private, secluded space of unknowable thoughts and deeds. Freud's interest lay in the fact that the word *unheimlich* is also used to convey something that is concealed and therefore potentially malevolent. He concludes: 'Thus *heimlich* is a word the meaning of which develops in the direction of ambivalence, until it finally coincides with its opposite, *unheimlich*.' The English word for *unheimlich* is 'uncanny', which in its Freudian sense, among other things, refers to strangely familiar experiences or, perhaps more horrifyingly, a sense of estrangement in what should be a familiar, homely situation. The upshot for Freud is that, like the Beast's palace, all homes are unhomely: cosy, talkative candlelight throws deep shadows, and for all the merry teatimes there are people withholding things from each other, unable to express the most meaningful parts of themselves.

Airbnb has set up shop in this uncanny valley. The guarded and enclosed aspect of homeliness diminishes with Airbnb's mandating of the inside-out house, the home whose rooms one can browse online, but even this transparency entails a certain uncanniness. As Freud remarks, 'everything is uncanny that ought to have remained hidden and secret, and yet comes to light'. I ought not to have known about that steeped urine! In the comments sections, the secrets of private life are broadcast, and strangers become familiar. Indeed, the whole enterprise is predicated on a classically uncanny oscillation between strangeness and familiarity. Airbnb invites its travellers to feel at home in the domestic space of a stranger, but in order for this situation to be tolerable, the uncanniness needs to be minimised. In other words, the stranger has to be converted into its opposite. Since in this paradigm being unknown is the same thing as being untrustworthy, the Airbnb website offers such assurances as 'We make it easy to get to know hosts like Michelle.' For her part, Michelle seems like an honest person, standing in her modern kitchen arranging daisies. One might wish to overlook the blur of three sharp knives stuck to a magnetic strip behind her left shoulder. Chesky's social vision certainly holds the stranger at knifepoint, since for all the chumminess there's a threat at the heart of his ethos. 'Some people', he says, 'will choose to be anonymous their whole life. That's okay. But if you don't opt into this online identity, you'll have less access to the services that require it. The rest of us build a history. We build a brand online.'

Some people will choose lifelong anonymity? What freaks! While not having much time for the naturally retiring among us, Chesky also seems to think that history and branding are the same thing. While branding often invokes 'tradition' – Mr Kipling wandering through a dappled orchard, or the peasant matriarch simmering blood-red pasta sauce – such commercial narratives are usually a smokescreen to the unlovely history of mass production. If brands are not, as Chesky implies, synonymous with history – if by history we mean what actually happened – then they're perhaps more aligned with what Chesky sees as a 'built' history. The manufactured nature of cyber-identities, while deemed vital to Airbnb's aims, is simultaneously an obstacle because of the connotations between manufacturing and illusion. What is to stop these selfie-brands from being dismantled and rebuilt in new shapes? How would you know it was me? For those who demand that the 4D body be as robust as its 3D precursor, there is always the threat of being duped by a dummy. As of 2014, Airbnb users in America were required to have their government ID scanned in an attempt to freeze the quicksilver out of their online selves. Responding to this move, Chesky said, 'We don't think you can be trusted in a place where you're anonymous.' For a pioneer trust-economist he seems wary of overestimating the scope of human integrity. This policy change is designed to intertwine the founding biometrics of citizenship with our brand image, composed of online displays of sanity: wholesome Facebook musings and non-violent tweets, scores of friends and followers, combining to make a thickly woven reed boat, whose density of woof and warp somehow assures the world that someone of substance is on board.

Do Not Be Afraid

While in the movie the 4D Man learns to ease his limbs through walls, in his 3D science career he is concurrently developing a super-dense, impenetrable metal called Cargonite, designed to fortify and contain. The film's irony is that he conjures with both liberation and confinement, and this double life resonates with the evolution of our own four-dimensional bodies. Digital technologies allow us to roam like never before, miraculously unfettered, pushing our way through the solid world, moving towards each other. But we are also exposed to a simultaneous drive for containment, which urges us never to forget ourselves as we navigate in this newfound everywhere. Here may have jumped into bed with There, but we must always be faithful to our digital brands. If we must pass through walls, we're asked to carve our signature into them as we go.

I wonder sometimes if it's worth having a four-dimensional body, only for it to be chased down and penned in. On days when the claustrophobia of tending to my online brand makes me regret all manner of digital connectivity, I try to remember Seamus Heaney. In the moments before he died in 2013, Heaney housed his last words in a text message to his wife, Marie. They were in Latin and said '*Noli timere*' – Do not be afraid. If you are grateful to Heaney's poems for fortifying you against fear, then it's easy to feel that everything necessary for that deathbed message to happen – the laying of fibre-optic lines, the erecting of the first cellular tower, Alexander Graham Bell gurgling in his cradle – was all leading up to the beautiful four-dimensional moment when Heaney tapped out these final letters. Thank God, you may think, after all he managed to communicate throughout his life, that this dimension was available to him and that he could reach his arm through all those walls to squeeze Marie's hand at the last.

Heaney's poem about the Irish hermit St Kevin offers a cellular vision of the ascetic, who is so tightly confined within his cave that, prefiguring an iconic pose of the digital age, he stretches out of the window with 'One turned-up palm'. But instead of the humid face of a BlackBerry, his hand connects with a blackbird, who 'settles down to nest'. By sheltering the bird St Kevin becomes 'linked / Into the network of eternal life' and 'Is moved to pity'. This linking-in arises from the interface between living matter: 'Kevin feels the warm eggs, the small breast, the tucked / Neat head and claws'. I'm reminded of the iPad granddaughter, who is apparently developing the digital nerves to feel the touch of her grandfather from hundreds of miles away. The four-dimensional body is growing warm. Do not be afraid. Twenty years before Heaney died, the first ever America Online Instant Message (AIM), a service that dominated internet chat in the 1990s, was sent from another husband to another wife. On 6 January 1993, Ted Leonsis wrote: 'Don't be scared . . . it is me. Love you and miss you.' The fear that needs consoling seems to be the shock of this new embodiment, occurring in the strange space somewhere between here and there, what Heaney refers to in the poem 'Electric Light' as 'The very "there-you-are-and-where-are-you?" / Of poetry itself.'

2

A Different Kind of Buzz

_____

Hive Life

You will likely have heard that there's a problem with the bees. They are on the wane – poisoned, many say, by crop sprays – though there seems to be no single cause for the epidemic known as Colony Collapse Disorder. An early hypothesis was that cell-phone signals were deranging them, silencing them with our talk. But the absence of a single culprit suggests that the bees are being overwhelmed by a range of environmental factors. Their deaths by the million testify to the multifaceted degradation of their world, which overlaps generously with ours. Honeybees are made fragile from being shipped around continents, their food supply is impoverished, their air more toxic with neonicotinoid pesticides. They are, in their own way, victims of globalisation. Exhaustion and malnourishment promotes the spread of pathogens in the hives. And so the sudden, swollen, undulating arrival of a full-blown bumble becomes one more thing lost to the past's sunlit gardens. Today, when one of their dwindling kind veers in the open window, I feel as in the presence of a time traveller, or revenant. I can surrender half an hour to watching this posthumous little creature nose and fret a useless stretch of glass. I love it and fear it; the swat has been relegated to the past as well. The swat belongs to the brutal, carefree 1990s. And what relief when the bee plunges, exhausted, and remembers the difference between window and air. Watching it fly vengelessly into the blue brings a sense of moral achievement.

Today, bees embody a stark pattern of sentimentality and apocalypse, past and future, a golden era alternating with the threat of dark days. The bass thrum of a bumblebee is the sound of vitality, since the vibrations that are the buzz's source disturb more of the flower's pollen onto the bee's body. Then off swerves another dusty, efficient progenitor of a third of our fruits and vegetables. No matter what legions of people might be writing in sympathetic emails, the bees are the ones really sending us good vibes. We seem to intuit this gift, since places alive with bee-buzz are archetypal Edens. The hum of bees worries the air into tranquillity. W. B. Yeats describes bees as a herald of peace in his poeticised dream of a remote cabin, on the isle of Innisfree, where he would 'live alone in the bee-loud glade'. But in their present state, bees are associated not with idyllic noise but with the terrible silence of vanished colonies. The herald has become the harbinger.

The rise of Web 2.0 roughly coincided with this global bee crisis, and thus a poetic quirk of the mid-2000s was that, while farmers were starting to report the disappearance of their colonies, we were becoming more bee-like. This comparison comes from a general eagerness to find metaphors for the digital revolution. The speed of our so-called online immigration, a metaphor in itself, is such

to a different kind of buzz. The story of the farmer kissing pollen-dipped chicken feathers against pear blossom is unsteadying for having a pre-industrial inefficiency caused by post-industrial reck-lessness. But it's also a potent contemporary image because it mirrors the prominent digital metaphor that figures online migration as a journey to a virtual hive. We are aping bees in multiple dimensions.

The old appetite to compare human civilisation to the life of bees is in some ways a remarkable one. We don't project ourselves so willingly onto our closer primate relations, likely because the uncanniness of the similarity is less than flattering. Bees have an undeniable charisma, inspiring in us a romance of ourselves. They are perhaps the most alien of all the animals we domesticate, and yet, from our point of view at least, there is a long-standing kinship. They share our taste for poetry, being the only known creatures beyond the primates to employ symbolic language to communicate. In their industrious, darting busyness, bees fill the hedgerows with a portable morality. According to the Jesuits, one can learn virtue simply by observing them. The beehive symbol has of course been put to imperial purposes. The ancient Egyptian elite saw the hive as a model of dynastic supremacy: masses of workers coordinating their efforts for the betterment of a single ruler. But the metaphor need not be tilted towards exploitation and subservience. The Masonic interest in bees centres on this notion of industry, of the possibilities of organised collective activity. Albert Mackey's nineteenth-century *Encyclopedia of Freemasonry* cites the beehive as 'the perfect emblem, or typical instance of the power of industry, because what no one bee or succession of separate bees could accomplish is easy where hundreds of them work together at one task at one time'. The scholar Juan Antonio Ramirez has traced the historical influence of bees on human architecture, and writes that 'the long period between the Napoleonic Wars and the Second World War perpetuated the metaphor of the beehive as the perfect united and industrious society.' Bees are thus integral to a certain aspect of our imagination. We look into their five eyes and, miraculously, see a perfected vision of human potential.

There's a black joke in us buzzing through cyberspace while the meadows fall quiet, as though an unspoken one-in-one-out policy is in effect. It is understandable, however, that a hyper-connected digital environment would be an especially friendly habitat for bee metaphors. The hive is certainly an intuitive image for the mass networking of human communication and the coordinated thought and activity that it makes possible. But the history of this metaphor reveals more about our current hive lives than merely our networked consciousnesses. And as we'll see, if we expand the metaphor beyond the idea of group thinking and industry, we find that our primary senses are also assuming an apian quality in their new four-dimensional habitat.

Wikipedia is a fine example of a hive endeavour, with hundreds of thousands of people working together to create an immense reference of human knowledge. The multimillion-dollar phenomenon of crowdsourcing, whereby a vast range of projects are communally financed using online platforms, is another incarnation, while digital technologies facilitate real-world swarms of protest and uprising. Social media, at its best, is in some sense a landscape of beehives in the clouds. A mostly digital friend of mine explicitly thanked the 'hive mind' after inviting it, via social media, to give him suggestions for good 'summer salads'. Feeling in a gastronomical rut, he was soon inundated with recipes. This developing conversation became for me a strange version of breaking news, procrastination's blend of the urgent and the inconsequential. It was a pleasant thread to weave through my lapses in focus. There was real joy in this exchange of ideas, and you could feel people's relief in this prosperous use of the technology. Commenters began to revel in the fact of the conversation itself – its usefulness and goodwill. They wrote happily among a bounty of fruit and vegetables. A spontaneous feeling of holiday emerged, of a rare occasion when the troublesome members of an extended, dysfunctional family have failed to turn up, leaving the nice ones to discuss the merits of spinach in peace.

As an outcome of hive behaviour, this camaraderie would, for those who distinguish between the hive-making creatures of the earth, symbolise the warmth of the bee community. The anthroposophist Rudolf Steiner, writing in the 1920s, saw the difference between bees and other group insects as lying with the animals' relationship to the goddess of love: 'If we describe the wasps and ants we can say they are creatures which, in a certain sense, withdraw from the influence of Venus, whereas the bees surrender themselves entirely to Venus, unfolding a life of love throughout the whole hive.' Isn't this a utopian vision for the Twittersphere, to submit to the influence of Venus? Much of our disappointment over social media occurs when it mimics the wrong sort of hive, one full of aggression, militancy and venom. The Twitter Storm is alas a more salient term than Twitter Sunshine. The prevailing dream for our online connectedness is that we collectively manage to assert and maintain a spirit of nourishing communality, which historically we have associated more with bees than with any other hive-dwelling creature. As Steiner puts it, albeit in the sexist language of the day: 'When one stands before a hive of bees one should say quite solemnly to oneself: "By way of the beehive the whole Cosmos enters man and makes him strong and able."'

Of course, we should always be vigilant as to the metaphors

we keep, especially the most romantic of them. There may be a less evident side to the comparison. The technologies of beekeeping, and the architectural and sociological visions they have inspired, have undergone a very digital progression from obscurity to visibility. The evolution of the domestic beehive has followed a trajectory from privacy to surveillance, one of the digital age's more sinister dilemmas. For thousands of years, the skep – that traditional straw basket with small holes for entrance and exit – served as the artificial home of the kept hive. The skep's upturned, domed habitation evokes a monastic stone cell. A revolution occurred in the mid-nineteenth century with L. L. Langstroth's design, the first to divide the hive into movable frames within a larger structure. These frames allowed keepers new access to inspect the health of the colony and to extract honey with innovative ease.

This trajectory is also present in the influence of bees on our buildings. Western architecture had for centuries been incorporating into its designs the hexagonal honeycomb formations and the parabolas found in natural hives. But then modernist designers combined this familiar inspiration with their avant-garde love of glass. The honeycomb logo of Peter Behrens's AEG turbine factory in Berlin makes the apian influence explicit. Built in 1909, the factory has a six-sided gable, which looks as though a hexagon has been cut and shaped into a six-sided semicircle, a honeycomb parabola. Long grids of windows run the length of the high stone walls. 'As an imaginary panopticon of labour,' Ramirez writes, 'anyone might consider this factory as a perfect society of workers, a genuine "[workers] observation beehive".' In the same period, German author Paul Scheerbart published *Glasarchitektur*, in relation to which Ramirez claims, 'It seems that in this hypothetical world of the future there is to be no hiding-place.' And yet a scrutinised beehive is anathema to the innate privacy of bees, whose very collectivity is a form of concealment. When constructing their sweet chambers the bees cluster thickly, their bodies forming a screen to their industry. They build their honeycombs unwatched, in darkness.

Nevertheless, the beehive was incorporated into utopian visions of a transparent future. The hexagons of the honeycomb inspired Le Corbusier's earliest skyscrapers, but we shouldn't forget that these skyscrapers were to be built of glass. The collectivism of the hive, in other words, became entwined with the modernist virtue of visibility. There is thus an apian echo in our present, digital debates about the ethics of transparency. One of our main contemporary anxieties is that digital life robs us of hiding places. An ever-normalising emotion of these times, a feeling that is fast approaching an assumption, is that our networked activity is being watched, that at any moment we can be lifted into the light like a bee on a tray. Here the bright side is the dark side. The buzzing exchanges of the online world may borrow metaphorically from the image of the hive, but it's significant that they have also been subjected to the ethos that would discipline the swarm with scrutiny.

While the metaphor of the digital hive speaks to our connectivity and the potential for group industry, our hive lives are making us resemble bees in another much more curious way, and one that affects us individually rather than collectively, at the level of the senses. The sensory physiology of bees is an ongoing mystery; only recently was it discovered that they can sense a flower's electrical field, which helps them to determine which flowers carry the most rewarding yield. In 1923, during a series of nine lectures on bees, Steiner proposed that bees navigate using 'a sense which is between taste and smell'. He described a worker-bee 'tasting' the sun and distant flowers, with 'no need to use its eyes at all'. Although Steiner's treatise was often inspired more by mysticism than empiricism, he does capture something of the bee's sensory complexity. Almost a century later, scientists can't always be sure if bees are smelling or tasting something, since the sites where they do their smelling and tasting are not, as they are in us, discrete and localised. An adult honeybee that leaves the hive to forage can both taste and smell with its antennae and forelegs, as well as its mouthparts. If we are going to pursue the digital bee analogy to its fullest ends, then we shouldn't stop with hives, but consider the sensory parallels. Whereas the two chemical senses of taste and smell are sometimes indistinguishable in bees, our own digital hive behaviour is producing an analogous blending of the physical senses of vision and hearing. Here we meet another notable structure of our rewired physiologies. This sensory blend is responsible for an aural paradox that is emerging in our times. For at the heart of four-dimensional sound is a riddle worthy of Gollum: What is becoming quieter and noisier? This nonsense only begins to be possible if we realign what it means to hear and to see, and indeed digital life is hacking these very ideas. As we turn to the notions of sight, sound and silence in four dimensions, I imagine Gollum's grim, crouching excitement: What creature has ears in its eyes, and mouths in its fingertips?

Soundproofing Silence

Because Shakespeare, it seems, knew everything, he knew that civilisation is constituted primarily through sound. Hamlet's famous last words – 'The rest is silence' – are less punning than ironic, since both his parting, eloquent gasps and his death play out amidst a growing bassline beat. 'What warlike noise is this?' Hamlet asks as the poison takes hold. The drums and commotion signal the arrival of the Norwegian crown prince Fortinbras, who bursts into the quiet of the massacred Danish court. From the beginning of *Hamlet*, we're taught to think of sovereignty as a manipulation of sound waves. The ghost of Hamlet's kingly father stalks the misty parapet complaining that his murder is an insult and injury to 'the whole ear of Denmark', which has been poisoned to the truth. It's understandable that poisoned ears are on the ghost's mind, given that his brother killed him in his sleep by pouring mercury into his own. The tragedy ends with Hamlet's corpse being carried away on a tide of noise and Fortinbras announcing that 'The soldiers' music and the rites of war / Speak loudly for him'. The play's last line is a stage direction, 'a peal of ordnance is shot off', which is another concluding pun because 'ordnance' is both an umbrella term for artillery and 'ordinance' is an authoritative order or command. These shots signal Norway's assumption of the kingdom, the ear of Denmark now listening to a resounding transferral of authority.

One of *Hamlet*'s messages, then, is that social organisation isn't a quiet affair. Think of the Greek agora, parliamentary Ayes and Nays, the tyrant's rousing balcony speech, the rebel's rallying cry, the pledges of allegiance, the royal anthems and the anarchists' songs. At worst you have edicts, at best you have debate. Democracy is designed to honour a plenitude of voices. As civilisations grow more complex they tend to become noisier, and indeed one of science fiction's preoccupations has been to prophesy the advanced stages of this garrulousness. When *Star Trek: The Next Generation* introduced the Borg to its roster of foes in the late 1980s, these hybrids of machine and flesh embodied all manner of *fin de siècle* anxieties over the endgame of technological progress. Their faces are half covered in black plating, from which thick wires extend and coil, some of which, in that oddly entrancing nightmare of subcutaneous circuitry, push through the flesh of the cheek or jaw. But the Borg were also unsettling for insinuating that, in a technologised future, silence would be relegated to the archaic past. Their bionic consciousness is processed through a unifying network called the Collective, into which they are permanently hooked. Theirs is a hive sensibility of the most unappealing kind. In one episode, the *Enterprise* crew rescues a lone survivor from a Borg shuttle that has crashed on a desert planet. This boy-Borg, as pathetic as a helpless teenage Nazi, remains psychologically deserted among the self-satisfied liberalism of twenty-fourth-century humanity and their similarly evolved posse of friendly aliens. He is homesick for the incessant aural companionship of the Collective, from which he has been stranded. 'Here it is quiet,' he glumly tells his captors. 'There are no other voices . . . on a Borg ship we live with the thoughts of the others in our minds. Thousands of others with us always.'

In our civilisation's advancing years, technological progress is, by all appearances, making life noisier. We can best discern how loud our lives feel in the moments when we try to discipline both real and virtual decibels. Remembrance Day is one of the few mainstream occasions to treat silence, in the old ascetic style, as a moral proposition. In the run-up to the 2012 tribute to the war dead, the Royal British Legion campaigned to extend the customary two-minute silence into cyberspace. This campaign alone attests to a collective recognition that online life is booming. Yet the irony of such a silencing venture was that, in order to reach a critical mass of online observers, the organisers first had to shout over the daily hubbub of Twitter and Facebook. This plea for quiet began as a harnessing and coordination of online noise. The British Legion became UK pioneers in the use of an American application called Thunderclap, which gathers subscribers to a single message – 'Cancer, we're coming to get you', 'We need #JobsNotCuts' – and then simultaneously releases this message en masse onto the social-media pages of all those who have signed up. The British Legion's mantra, 'I'll be remembering the fallen at 11 o'clock #2MinuteSilence #LestWeForget', thundered across Twitter timelines on the morning of Remembrance Day. Owing to the six-degrees-of-separation of social networking, over 10 million users were alerted by 20,000 people hitting their forks against their glasses at the same time.

Having survived both the roar of the Thunderclap and the ensuing silence, poppy-adorned Twitter solemnly raised its head and allowed the flow of un-solemn thoughts to resume in its vast, collective mind. Soon, aspersions were cast on those in breach of the two-minute reverence. The presenter Jeremy Clarkson, who has built a career on his indignation, was quick to voice disapproval when an automated tweet advertising his own programme *Top Gear* broke the Twitter pause. Joining a train of outraged tweeters speaking of 'shame' and 'disgust', Clarkson wrote his own chastising tweet, calling the mistake 'astonishingly stupid'. This quick-tempered policing of silence, and the difficulty of achieving even a strange simulation of it, reveals what an exotic state silence has become. And yet one of the weird aspects of our times is that while silence retreats we are simultaneously becoming quieter. We say less to one another out loud – as of 2012 in the UK, text-messaging officially eclipsed both phone calls and speaking face to face as the main form of daily communication.

At this point we might think of beekeepers tapping the sides of their hives, listening for the reassuring hum that never comes. Almost as if in sympathetic muteness with the world's collapsing bee colonies, our own hive lives prosper in silence.

If the digital age were straightforward and in need of a pantheon, we might be tempted to say that Thunderclap is reminiscent of Zeus and his disciplinary lightning bolt. Homer, after all, calls him 'the Thunderer' and 'the Cloud-gatherer'. In the *Iliad*, when Zeus warns the other gods not to involve themselves in the mortals' war, Homer tells us that 'they all fell silent as he spoke, astonished by the force of his words'. When he 'thundered aloud', hurling a bolt among the Greek army, 'they were dumbfounded'. Here Zeus' use of noise to shut people up is orthodox and by-the-books. But digital life, particularly when it comes to sound waves, has a kink to it. It deals in silent booms of thunder; so much of its noise, so evident and so *heard*, is in reality an absence of noise, a cacophony of mouthless text and pictures. Indeed, with its present form being such a predominantly visual medium, digitisation gives itself a more complete sensory range by way of metaphor: consider, for instance, the photographic meaning of 'digital noise' as referring to a graininess or unwanted pixellation of the computerised image. Thus the god for whom we're looking isn't Zeus but Dionysus, born, like an accidental text message, of Zeus' thigh. Dionysus' sense of aural kink matches this digital perversity. Like his father he has deafening monikers: 'the Roarer' or 'the Loud-Shouter'. His trademark entrance onto an Arcadian scene involves blaring pandemonium, but crucially it is from within this noisy ecstasy that a terrible silence petrifies any mortal present. As Walter Otto wrote of his dreadful roar, 'it is as if the insane din were in reality the profoundest of silences'.

This slipperiness between noise and silence makes digital life Dionysian. Collectively we invoke the bellowing god. To create this communal sonic boom in cyberspace, we ironically fall quiet, our fingers gliding over touchscreens, our chins pleated. I recently teased one of my young American friends that she and her entire generation find everything awkward. 'I *know*!' she said. 'It's awful. People can't even stand to be silent around each other any more. Everything is awkward.' It's true that the term 'awkward silence' seems to have dropped out of usage, precisely because it has become a tautology: all silences are awkward. Perhaps most awkward about these modern social silences is that they are often double-layered, consisting of the silence between those gathered stiffly together and the private silences into which they are likely to retreat. There was a time when the pressures of quietness would generally force us into some casual, throwaway noise-making, but now we fight like with like, and only barely get away with it. Increasingly our standard defence against the social unease formerly known as the awkward silence is not to make noise but to stroke our little talismans of talkative silences. We outsource our incidental chat, idly checking email or Facebook or sending a text, hoping that the fleshy aggregation loitering in soft focus beyond the crisp horizon of our smartphone will soon move on. We fill 3D silence not with 3D small talk but with our respective 4D silences, our noisy chaperones.

Indeed, 4D silence has turned up the volume of life so much that even our eyes need earplugs. This sensory delocalisation evokes once again the honeybee, with its widespread capacity for smell and taste. Most significant about Clarkson's irritation is that it betrays the extent to which we're now prone to think of online text as possessing an audible dimension. It is becoming intuitive to think of silent images and strings of text as forms of noise. In the case of the *Top Gear* tweet, the shameful opposite of silence wasn't noise but the invitation to distraction. With this expansion of our experience of noise, it will soon be necessary to broaden the definition of the word 'mute', which is certainly in the ascendant. Mute may end up being the verb of the century. I remember noticing the mutation of muteness while trying to get rid of an unwanted advert crowding in on my YouTube viewing. It was one of those online billboards, an inert, soundless, digital poster. Moving my cursor pitilessly towards the X button, a piece of rollover text intervened like a social worker, asking if I wanted to 'Mute This Ad'. I did so, and a square of plain colour covered the billboard, displaying the words: 'Ad Muted'. The ad re-formed when I clicked again on the Mute banner, loitering conspicuously once again beside *The Hollywood Reporter* Actress Roundtable.

Google launched the muting option for its ads in mid-2012. Nowhere in its history has the word 'mute' been related to obscuring or rendering invisible; it has always been associated with the absence or disciplining of sound. Increasingly we're trying to soundproof the soundless. Noise-dampening applications, which allow Twitter users to silence their followers without deleting them, are coming to seem like a necessary aspect of online life, a sign of our relentless proximity and its strains. In May 2014, Twitter announced its official silencing feature. Note the cross-wired senses: 'Muting a user on Twitter means their Tweets and Retweets will no longer be visible in your home timeline'.

It is of course still possible in digital terms for mute to mean the silencing of the audible. In the video game *Grand Theft Auto*, users are aggravated by the actual, old-fashioned noisiness of their real-life, disembodied opponents, with whom they interact over

suddenly goes dead. The last comment dangles above silence, made foolish somehow, unless it has the glossy finish of a concluding remark. It can feel as though one member of the conversation has been carried off by aliens, mid-sentence. This is partly caused by social media's hybridity, its crossing of conversation with note-writing, the chess-game-by-post of intercourse. The form is naturally suspenseful, and it can be unclear when the end of an exchange has come. As a result, I've gained a new appreciation for the phone call's conclusive rat-a-tat of goodbyes, or for the gentle and courtly rituals of fleshly farewells, the steps backward and forward, the inane but consoling waffle, the promises of future reunions. We naturally sweeten partings with noise. On social media, it's easy for the silences endured by others to sting us mildly as we scroll through the day, but these stings are surely good for our empathy. I'm thinking in particular now of two friends, each of whom I know better than they know each other, having a rare online exchange. After some specific business, one of them broached the idea of a coffee. I came sniffing around these parts hours later, and felt a pang for the friend whose invitation was swaying bleakly at the end of its tether. But then I saw that the invitee had 'liked' the proposition, a non-committal but humane closure, a little punch of noise to plug the silence.

We've become such shrewd listeners of those digital affirmations. We wait with beating hearts as a thread is pulled taut with social tension, and breathe easy when irreconcilable moral differences are breached with mutual thumbs aloft. What relief, what diplomacy, when the butt likes the joke. Social media is a watchful medium, but that watchfulness combined with its busyness trains us, in the logic of blended senses, to have absolute pitch for the various notes of silence. The garrulousness makes us alert to those things that have gone unsaid. To withhold one's seal of approval is a political statement, and even in spied conversations between total strangers we can read patterns in these silences, in the spaces between clusters of endorsement. If a comment or picture has three likes, it can be fascinating to scroll nosily over, to see if the roster of names matches your predictions, to assess the implications of missing signatories. 'She's never once retweeted me!' a friend complained of a burgeoning acquaintance, and relations have since curdled in this dumbness. And because in this dimension the visual is blended with the aural, the lack of an anticipated picture speaks volumes. That which Instagram leaves fully in the shadows, unseen, unshaded and undappled, voices much about favour and its opposite. We hear stories in the silences, in absence's barbed kind of presence.

This watchfulness and careful listening for cues is a continuation of thousands of years of social life, with its numerous species of snub, but our online noisiness makes our silences more stark and evident than ever before. At all times of day we can see without being told that our silent partners are chatting away merrily to others, which can create a peculiarly relentless sort of misery. A Facebook message's 'seen by' function is a recipe for hurt feeling. It's hard not to take silence personally when its source is blaring elsewhere. From the angle of the sender, silence, rather than sound, radiates from this vision. In the digital world, if seeing is hearing, then seeing seeing can amount to a dismal not-hearing. Such a culture, furthermore, makes you aware of how loud you're being. If I have long owed someone a message and stubbornly refuse to alleviate myself of this guilt by replying, then I feel compelled to tiptoe around the rest of the network. Even an unthinking 'like' can become a creaking floorboard. The temptation is to stay very still, very quiet. One silence can in this way make a hundred others, out of politeness.

The Silence of Marcel Duchamp

Writing about manners and mores in the 1940s, the critic Lionel Trilling praised novelists who could transmit in their works 'a culture's hum and buzz of implication'. Novelists concerned with the present will have to tackle the exceedingly complex buzz of our times. For on one hand, our present is characterised by a frightening and distinctive non-buzz, the silent alarm of our environmental mismanagement. On the other, we grow ever louder, except that this loudness is its own kind of silence, and indeed all manner of actual silences proliferate in the gaps between this noiseless hum. The paradoxical, Dionysian quality of four-dimensional sound – the roar that is also a terrible silence – is even more apposite here if we note that Dionysus was an early keeper of bees. In the myth, the satyrs in Dionysus' train were making an uproarious noise, crashing their cymbals together. This carry-on attracted great swarms of bees, which Dionysus rounded up into a hollow tree, wherein, shortly thereafter, he found honey. Bees thus busy themselves throughout this story of four-dimensional thought and sensation, in both the industrious, coordinated groups of social media and singly, in senses mingling across the body, in digitisation's synaesthesia of sight and sound. They are in loud abundance everywhere, in fact, except where they are needed most.

Contemporary novelists interested in timeliness might consider a figure from the recent past, another apparent time traveller, who was, it seems, tuned into our current frequencies. For an aspect of the artist Marcel Duchamp's genius was his ability, in Trilling's terms, to transmit the hum and buzz of the future. Duchamp's strange later life, which the people around him described as a sort of afterlife, symbolises our era's interplay between silence and sound, as well as producing as its legacy an artwork that whispers intensely to our digital times. Duchamp was certainly out of tune with his own age in the sense that he didn't share other modernists' hopes for utopian collectivism. 'I don't believe in the ant-hill society of the future,' he said, nor presumably in any other variety of hive: 'Every man for himself, like in a shipwreck.' From the mid-1920s until his death in 1968, he was seen to have abandoned art-making. He claimed to have become repulsed by his own industriousness, by the fact that, should he want to, he could make three of his coveted ready-mades per day. He would figure his obsession with chess as a means of thwarting his productivity. His withdrawal from the art world became, perversely, his conspicuous mid and late style. One friend said of Duchamp that 'In a way he gave up life while still alive.' Another saw him as exemplifying 'total physical detachment'. In 1964, Joseph Beuys gave a live performance piece on German television that included his assistants writing on some paper the sentence 'The silence of Marcel Duchamp is overrated.' The cult of renunciation surrounding Duchamp agitated Beuys, who remarked that Duchamp 'wanted to become a hero in silence'. Beuys, who expressed his enduring fascination with the apian community in works such as *Honeypump at the Workplace* and *From the Life of the Bees*, once said that 'only cooperation exists'. He believed that a successful society is one in which Eros itself is collectivised. 'All in all,' he maintained, late in his life, 'socialism is love.' Duchamp, who in the 1920s cultivated a transvestite alter ego Rrose Sélavy (*Éros, c'est la vie*), seemed to prefer, as many of his cousin surrealists did, highly individualised, incommunicable eroticism. Love as shipwreck.

Yet Duchamp's silence was ultimately a muffler for something

From the Communist point of view, capitalism and four-dimensional life share certain similarities. The *Manifesto* claims of the former that 'It must nestle everywhere, settle everywhere, establish connexions everywhere.' Few could object to 'all that is solid melts into air' being the tag line to the digital revolution, which has managed to combine, in a steamy double whammy, the neo-liberal smelter with the vaporous outputs of digitised media. For capitalism has indeed found a willing host in the light, airy realm of the digital. If we forget for a moment, as we do most of the time, about the massive racks of servers in data centres around the world, forever threatened with overheating as they preserve the cloudiness of the Cloud, then light and air are the native media of our digital experiences. The immateriality of digital life has always been one of its explicit promises, evident in the dream of the paperless office, or the queer possibility of bookcase after bookcase dissolving in the liquid screen of an e-reader.

While we knew from the start that the revolution would dematerialise all sorts of objects, we didn't wholly foresee the interpersonal fallout. The ability to coordinate virtually and efficiently using digital means has vaporised some charming, incidental figures from our lives. To return to our earlier theme of holiday lets, a topic apparently on which I often brood, the last few times I've rented a short-stay cottage I've been struck by the imma-teriality of the transaction. I can go from booking to considerately stripping the bed without seeing or even speaking to anyone involved in the enterprise. The money beams from one virtual pot to another; only the house, chosen online from a selection of intangible images, eventually materialises. During one outward journey I exchanged some pleasant emails about snow and bedding with my host while we were both on trains, and on the final morning I texted her for check-out times. But all in all her solidity, her personal touch, amounted to some thoughtful teabags, a bottle of wine and a packet of white-chocolate-and-strawberry hearts, none of which, I should add, danced or sang or whispered in corners. Seldom in these times do you pick up the heavy iron key at the farmhouse from the kindly farmer's wife, where over a glass of milk she dispenses the advice typically housed in that binder you find now on arrival, sitting impersonally on the glossy, unfamiliar tablecloth. At the farmhouse you get the instructions but a lot else besides: the farmer's wife's broad, lined forehead, her apron marbled with wholesome spills and wipes, the drowsy cat stretching its legs, the scuttle of black eggs by the coke fire. We get all that nineteenth-century realism, the experience of that extra solidity, as compensation for less-oiled mechanisms of trade.

I'm not advocating that the farmer's wife stay locked in place to satisfy my fascist romance of the countryside. I'm glad she's out and about on the East Midlands line, and able to keep the home fires burning from her smartphone. I conjure this rustic phantom of her simply to show how the opportunity for ghostly transactions has intensified in these digital times. Moreover, of real significance is that this ghostliness is only possible, as discussed in the materialisation of the four-dimensional body, because we have solidified elsewhere, in our online manifestation, which can be tracked down and called to account. There's no getaway car from the getaway cottage. If we can move invisibly through the real world, it is because the uniform of digital life includes a high-vis jacket. The dynamic equilibrium between ethereality and substance, like a balanced chemical equation, is a fascinating quality of the four-dimensional experience. For the unexpected thing about digital technology isn't so much its immateriality, which has always been a principal selling point, but rather its way of solidifying the intangible. All that is fluid sets into stone.

This solidification arises from those activities that digitisation makes easy – sharing, repetition, the galvanising of mass frames of collective reference – and which entice us to partake in one of digitisation's main interests: the interpretation of the world in terms of organised sets of data. We're learning to talk about life more as a collection of objects than as a sequence of actions, a privileging of monument over motion. The communality of digital culture encourages us to sort the swing and tramp of life into a series of types, and to talk to one another in these terms. The four-dimensional human, I'll argue, is perpetually training to be a taxonomist. This rage for categorisation has definite repercussions for the sense of our own originality, and, by extension, the possibilities for individual style. How do we dress for the fourth dimension?

In this milieu of solidified ether, one school-day tradition appears to me now like a sort of allegorical tale from the deep, pre-digital past. Each December it came time again to place your order for the Candy Cane charity drive. Typically it was an upper-year girl, who in my younger days seemed as ancient and wise as the hills, who would staff the makeshift stall by the water fountain. The gravitas of her office gave the fold-out chair and errant class-room table an air of civic authority. At break times she would permit you to approach her. For twenty-five cents (five for a dollar), you could request a candy cane for a friend and write a short note on a stub of rough paper, the memory of which, in later life, is fated to return whenever a dry cleaner hands over a ticket. Then at the last lunch of the term, between the turkey and dismissal, a squad of upper-years would march curtly among the diners, depositing the long-awaited treats.

And so, inevitably, there was a sudden, striped manifestation of the pecking order, popularity confirmed in the glistening, crackling bushels bound in rubber bands and dropped plate-side. With practised resignation, those less fortunate would have to help hoist yet another of these stuffed, miniature umbrella-stands down to the cool end of the table. Meanwhile, the most socially wretched would limp out of the dining hall with the lone cane from their pitying form tutor. As a mid-level nerd with a modest collection, I found the whole spectacle a morose but blessedly temporary crystallisation of 'where you stood'. At least the year was over and there was nothing but the carol service left to survive, at the end of which the customary shredding of the hymn books during the last song would culminate in a democratic snowfall of paper. This confetti, if not a wealth of candy canes, would be general over the gymnasium, made by and drifting over the winners and the losers.

Mercifully, the Candygram's tactless quantifying of social success was a vivid rarity, a pageant of glory and mortification that lasted a lunch hour and which published the raw data of dorkdom. School politics and apartheids are depressing enough to witness in their implicit guises, without these rock-solid markers. But the current format of social media has made every day an extravaganza of candy canes. Every picture, post, upload, comment, or comment within comment, sends the prefects doling out the ticks. For the young especially, and no matter the platform, the glaring total of social-media allies must feel stitched onto the breast pocket. With so much of online conviviality expressed in the form of endorsements, once-tacit interpersonal dynamics are materialising. The abstraction of social grace is made solid, as measurable as bone density.

The allegory of the candy canes applies to the wider values of online life, which has been organised so that we can seldom navigate cyberspace without giving weight to things. Our online activity inevitably brings stuff into being, into solidity, from our clicks and views and general involvement in the circulation of information. Every wander among websites adds to the heft of their 'analytics'. Put another way, we are unwittingly involved in a game of pass the parcel where the parcel gets bigger with each handoff, another layer of wrapping added rather than ripped away. Power gathers around sections of networks that have grown fat and heavy with hits. In this immaterial environment, weightiness is revered. Klout, the controversial analytics company, which calculates a 'social influence score' for individuals and businesses, as well as offering ways to improve it, is an unabashed disseminator of the idea that digital solidity is one of the new virtues. Solidity's cousins are influence, impact, sway. A clout is, after all, 'a heavy blow'.

When Marx looked around modernity and saw the technologies designed to reduce labour producing more work for the worker, and methods of moneymaking becoming causes of poverty, and unimaginable progress generating decay and horror, he realised that 'In our days, everything seems pregnant with its contrary.' He would no doubt have been alert to the particular interplay between evaporation and solidification in our times. In the belly of our virtuality is the impulse to solidify, to categorise, to name. Digital life, for all its ethereality, is in the business of making things matter.

The Noun Invasion

It's no accident that, more or less at the moment when life became markedly immaterial, people started asking: 'Is that a thing?' You may have done so yourself, and felt a little shy at first, and glanced around to see if you'd pulled it off. Suddenly, all sorts of phenomena are thought of as things rather than, say, happenings. 'Replacing the milk in your coffee with a stick of butter – is that a thing?' '"Virtual Girlfriends" are a thing now', reports a *Telegraph* headline to a story about people subscribing to online companionship-providers. This ontological craze spells a blow for eccentricity: 'Wearing one leather shoe and one running shoe? That's *not* a thing . . .' Things can be judged once they are indeed things; non-things aren't worthy of sustained attention. Actions are turned into tangible things through sheer repetition, like calling the Candyman five times in the mirror. This process is so marked in our times because repetition is one of digitisation's fortes – the whole idea of the viral video is its prolific repetition through cyberspace. What's more, acknowledgement of 'things' is a linguistic recognition of our collectivity, a way of summoning up the chorus. Every time we mention 'a thing', we tacitly invoke the multitudes who made it. A thing must be performed by the many in order to take form, to prompt a weird grammatical trick whereby verbs are petrified into nouns.

The verbs may have had it coming, since in the old 3D world nouns have taken a beating. As CDs, DVDs, envelopes, takeaway menus, atlases and pocket-sized language books rain down into dumpsters, we're looking around for the objects that might replace them. Science has known for centuries that matter can't be destroyed, only rearranged, and so it's apt that the virtualising of reality comes hand in hand with a concretising of experience. In a rhetorical attempt to honour the law of conservation of mass, there has been an invasion of nouns. The campaign to objectify the world can be seen in a certain species of status update that has enjoyed a remarkable heyday. This style is more easily shown by example than description, but you know the sort – it has definitely been a thing since 2012:

> When the person sitting next to you on the 33 bus at 6:15 on a drizzly evening presses a finger to their left nostril and blows a torrent of snot to the floor, barely missing the rims of the red-stitched Doc Martens that your ex got you when you had your wisdom teeth out. That.

Students of grammar will notice that this breed shivers with verbs, but none of them is the main one. They are all subject and no predicate, one enormous, trembling *thing*. A central villain of my timeline works prolifically in this genre, so I was never predisposed to it. But besides my particular ill will, something more generally significant about this craze is its own play between the particular and the general. These updates pass off the idiosyncratic as typical, turning the personal and specific moments of an individual's consciousness into general categories of experience. This irony is the spine of their style. Social media is an incubator for all sorts of irony, from the witty to the woeful, with the humblebrag – the falsely modest announcement – being the goblin monarch of the latter. The joke of these mega-nouns is their portrayal of life, not as a cascade of unrepeatable circumstances, but rather as a sequence of prefabricated scenes. Nothing is personal; everything is a type.

style verbs out of the sentence altogether. These are the *anti-gerunds*, the nouning of verbs instead of the verbing of nouns. For instance, in *The Mindy Project*, when Mindy comes round from general anaesthetic her gross employee Morgan admits to having repeatedly tried to kiss her awake: 'I've been *Snow White*-ing you for an hour.' Elsewhere, Mindy comments on the dangers of online identity fraud: 'I was *Catfish*-ed before there was *Catfish*.' In the buddy show *Broad City*, Abbi is startled when her friend Ilana appears behind her, suddenly visible in the mirror. 'Don't *Black Swan* me like that!' Abbi says. In *30 Rock*'s high-school reunion episode, Liz Lemon's former classmates plan 'to *Carrie*' her with a teetering can of red paint. If you belong to the precarious middle-class generation for whom staircases and home ownership are the stuff of childhood memory, then you and your beloved may be tempted to *Wings of the Dove* any peaky heiress who crosses your path.

This trend is a form of camp, which has always revelled in narrating one's humble life in iconic, cinematic terms: 'That's *so Some Like It Hot*' or 'Crisp white blouse? Very *Roman Holiday*.' In these cases, take note, the nouns are turned into adjectives, which is a less radical immobilising spell than the next phase of verbal petrifaction, since neither nouns nor adjectives move around in the first place. Theoretically, the older a society grows, assuming that it remembers the highlights of its past, the possibilities for camp become more numerous, and the more its language will be decorated with these noun-verbs. Victorians couldn't liken something to Mickey Mouse; they couldn't Peter Pan anyone. And likewise our twentieth-century selves were metaphorically poorer than they are now, unable as they were to compare anything to *The X-Factor*. Moreover, the shared reference frame that our connectivity provides means that it will be increasingly easier to employ such nominal shorthands. The more 'things' forged in the digital fire, the more we will be able to put them to use.

The endgame of this fever for categorisation will see the nouns inherit the earth. Everyday speech will be packed to the gunwales with allusion, a hoarder's paradise of intransitive, proper verbs. In the future, you won't get on a bus, go to the shops, buy a gift for your clandestine lover, bring it home, get found out and feel terrible: you'll *Love Actually* your long-suffering partner. All those boisterous actions, summoning movement and spaciousness, will be pinned under analogy. 'Love is a verb', sang Massive Attack once upon a time, but is *Love Actually* a doing word? On one hand there's the potential for great poetic flourish in this language of the future, which may blaze with colour and rely on the mobility of metaphor for its dynamism. While this caddish story may not send your listeners on a journey to your local shopping precinct, the economical, allusive version could deposit them instead in Emma Thompson's bedroom, exposed to an ocean of feeling in the simple readjustment of a bedspread. On the other hand, we will have to become accustomed to the sense that life is a collection of likenesses, lived out elsewhere, and that any forward momentum we possess is just the aftershock of all the things that have already happened.

A vogue for the steady footing of categories has in the past been one sign of anxiety over the speed of social change. In nineteenth-century Paris, when metropolitan life was young, newspaper supplements called physiologies were a craze. These generally mass-produced booklets included humorous sketches of the sorts of people whose paths you were likely to cross. They were an expanded, urban relative of those illustrated plaques in nature reserves describing a cornucopia of hypothetical ducks. The physiologies assumed that they could explain the city to itself, listing the dress and habits of the poet, the woman of the demi-monde, the *flâneur*, the provincial in Paris. In *The Physiology of Physiologies*, the author writes: 'Thanks to these little books, forged from science and wisdom, man will be better classified, better divided and subdivided than his fellow creatures.' Balzac added to this fad. In his 1841 *Physiology of the Employee*, he describes the many varieties of workers in Paris, the ordinary and the remarkable, who take after 'the plant and the animal, the mollusc and the bee'. He includes here the brutal axiom: 'The provincial employee is *someone*, whereas the Parisian employee is *something*.' The female concierge, we're told in James Rousseau's physiology, is a modern creation who, having yet to be replaced by a binder and a bag of chocolate hearts, can 'according to her whims, give you the good life or cast you into a premature hell'. No matter how satirical, by offering some ontological certainty in a swelling crowd of strangers, these caricatures categorised the populace.

The spirit of the physiologies persists today in our instinct for character types, except that in some corners we have even evolved beyond giving names to them. Today it's not so much a matter of describing the various species of person because, from the way we talk, it's apparent that we already know their particulars. *The Complete Physiologies* is now one of the free texts we all come with. For we no longer have 'the creditor', 'the gourmand', 'the shop-girl'; we have *that girl*. We handle these title-less types with tongs, usually to make clear that we share no similarities with them. 'I didn't wanna be that guy,' some might say, or 'I was going to complain that I ordered a gluten-free meal for the flight, but then I'd be *that* girl, you know?'

In 2013, country-music artist Jennifer Nettles sang about this very problem. 'Even though he's being that guy / I don't want to be that girl', the narrator confesses sororally to a woman whose 'man' is trying to get with her down an alley. Both the single and Nettles's subsequent album were called 'That Girl', while during the Y2K mysticism of the late 1990s and early 2000s the albums to which she contributed were named things like *Hallelujah* or *Story of Your Bones*. It would be hard to imagine this sense of 'That Girl' arising from anything but widespread online culture. The phrase itself carries the assumptions that purveyors of social media coax us into imagining, such as that we're all on the same page, and that we all have near-identical ranges of reference. It speaks to a deep and often cynical familiarity with the permutations of modern behaviour that may not even be genuine, and yet which somehow feels true as a side effect of connectivity.

Justin Timberlake also debuted a song in 2013 called 'That Girl', but his take on the phrase is old-fashioned, being of the same ilk as Stevie Wonder's song of the same name from 1981. In both of these love-struck ballads, the whole point is that the Girl is a one-of-a-kind, the one-and-only, not a category. Neither lad can resist calling out to her, in all her specificity. Wonder reminds That Girl about her responsibilities: 'I've been holding for a long time / And you've been running for a long time / It's time to do what we have to do'. For his part, Timberlake 'Didn't have to run, I knew it was love from a mile away / But I had to catch you, running through my mind all day baby'. Is it just me, or are Timberlake and Wonder being *that guy*? Run, Lola, run!

Each of these 'thats' comes with a tacit, world-weary sigh, and as is natural for new slang you hear them most from the young. Those of the luminous skin (*that* skin), which, even if you never had it yourself, you mourn for as a lost love, are the fresh-faced majority inhabiting this seen-it-all mood. Surprise doesn't come easily when the planet is overcrowded with precedents. This vein of twenty-first-century speech is a new linguistic style made to express a feeling of the present's staleness. As such, it might be called the slang of exhaustion.

Curiously, the vagueness of this new shorthand implies how precisely subdivided we now imagine the world to be. These types are so familiar, so present in all our minds, that we only need gesture to them. The 'That Girl/Guy' formula also suggests that there are now limitless ways not to be merely yourself. How do you be original, when all sorts of misdeeds turn you into a type? In this sense our four-dimensional selves, being consumers in an emerging market of trust, have an unsteadying task. While our digital anatomies are requested always to be recognisable, to be locatable in space and time and defined by friendship history and past activity, the same environment that demands this readily identifiable body also challenges the very possibility of idiosyncrasy. Individuality in a world of categories is a high-wire act: one false step and you plunge into the canyon of *those* boys and girls who have the misfortune of being both unappealing and generic. Here then we find the chain-store self in its full mediocrity, a franchise forced to bear the stylelessness of the no-name brand.

Remaking the Original

Adam was, by some accounts, the first great namer: 'The man gave names to all the cattle, and to the birds of the sky, and to every beast of the field'. So Adam got there first, and given the credible reports that we're living through the sixth great extinction, it's wise that we're finding other things besides creatures to name. Indeed, it is a melancholy thought that, while earthly species are dying out at accelerated rates, we are diligently recording the explosive speciation of our millennial mores. This noun invasion, as we've seen, can take as collateral damage the pioneer's zeal, the spirit of possibility. A verbless world is necessarily immobilised. There are of course the pleasures and ingenuity of naming, but, as Adam's story hints, this is artistry of the second rank. The beasts of the air and land had to be made before they could be named. The real creative genius had, in other words, already been unleashed by the great Originator. You can better understand the desire to smash Paradise when you consider how sewn-up it was. Before we pity Adam, however, we should remember that at least he had the entire, anonymous universe to tag. For us four-dimensional moderns, the nouns are so over-named that we've had to move on to the verbs. What's more, in the naming game Adam had no rival, whereas we can all be namers now, we are all involved in making *things* happen. Everybody's a taxonomist, flicking through index cards.

A digital environment, then, in producing a culture of taxonomy also creates a problem of originality. As one might expect, an age of information is not hospitable to certain types of illusion, one of which being the idiosyncrasy of thought. It may seem perverse to suggest, especially in a book about the dazzling newness of digitised personhood, that nothing innovative is going on. On the contrary, there now exists a Japanese robot called Pepper who can read our emotions, to give just one novel balm for loneliness. But our times are strange for being a friend to innovation and a menace to originality.

As an example of this quirk, let's take the Google Car, a thing that could not be accused, in the grand scheme, of being passé. Rumours that the automobile was in the process of being googled began a few years before it arrived. Then, in 2014, we saw a preview of this 'self-driving' car, a term that accelerates the issue of mechanical sentience more aggressively than, for example, the self-cleaning oven. With cars having gained a self, commentators speculated on what had been lost. Such laments have a petulance to match the Google Car prototype's childish design, which, as many observed, shrank the world to the scale of the toy box. The fretful should remember that in this one concession to a motorised self, hundreds of thousands of souls per year may be saved.

The freedom to worry less about loved ones being mashed in car accidents is somehow less vivid than our anticipated nostalgia for free movement, for untracked, autonomous driving. With the 1990s now squarely in our rear-view mirror – a wacky race featuring the Central Perk couch, a giant Fruitopia bottle and Two Fat Ladies – this decade has at last become nostalgia's province. The Google Car seems to be driving us firmly away from the 1990s, the last period to contain at least some pre-digital innocence. It's perhaps for this reason that the Google Car's unveiling has made people think back to the motoring icons of that time. For who could be more at odds with Google's progress than two different ladies, Thelma and Louise, who refused to believe in the end of the road?

On 30 May 2014, Elizabeth Renzetti wrote an article about the Google Car for the *Globe and Mail*, which began:

I am trying to imagine a remake of *Thelma & Louise*, set some time in the near future. Seated in their car on the rim of the Grand

Canyon, surrounded by police, with their only choices prison or freedom by annihilation, Thelma turns to Louise in the seat next to her and says, 'Just go.'

'I can't go,' says Louise.

'What do you mean, you can't go?'

'This thing doesn't have a gas pedal. Or a steering wheel.'

A few weeks later, Kevin Maney in the *Independent* launched his response to the Google Car as follows:

Imagine a *Thelma & Louise* remake, circa 2030. Climax of the movie: two women sit in a convertible facing the edge of the Grand Canyon. Police surge towards them from behind.

Louise looks at the dashboard. 'OK, Google Car – go!'

The car does nothing. The police close in. A disembodied voice chirps from the car speakers: 'I'm sorry, it is unsafe to proceed.'

'Damn autonomous cars!' Thelma yells as she pulls a revolver and shoots the dashboard.

By early July we still weren't finished, with the *New York Times*'s Delia Ephron asking us, similarly, to 'Suppose Thelma and Louise were on the run in the Google Car. Louise is not behind the wheel. There is no wheel. No Thunderbird. No top down . . . No, don't imagine any of this, it's too depressing.'

It is bad manners to collide opinion columns in this way, and if I'm lashing out it's because on first hearing of the Google Car I had made a mental note, softly haloed with the calm, steady light of my genius, which if transcribed might read: 'Google Car cf. T&L'. Indeed, I only know about these echoing exhortations by journalists 'to imagine' because I had googled 'Thelma & Louise

Google Car' and hoped for the best. Despite being the first in my unofficial ledger, Elizabeth Renzetti, unlike me, had no illusions about the originality of her thought experiment: 'I could not have been the only one picturing those road-trip desperadoes while watching Google unveil its postdriver car this week.'

Renzetti's concession will become a standard editorial hedge of the digital age, as it becomes easier to contain every writer with the same thought inside a web search. Google's domestication of the open road seems part of the same process as its pedantic exposing of supposedly wild ideas. For just as the road has already been mapped, the car's mind surging ahead of its wheels, Google – our great taxonomist – is also available to tell all critics but one that their imaginative leap into the unknown has already been made.

In Alan Bennett's *The History Boys*, Hector tells his students:

> The best moments in reading are when you come across something
> – a thought, a feeling, a way of looking at things – which you had
> thought special and particular to you. And now, here it is, set down
> by someone else, a person you have never met, someone even who
> is long dead. And it is as if a hand has come out and taken yours.

This humane sentiment is the triumph of community over originality. Here the question of ownership and indeed the very idea of selfhood dissolve. In the solitude of reading, two people are harmonised across space and time by a shared idea. But while Bennett's ideal vision is generally true for the reader, it's the stuff of nightmares for the writer. The reader discovers sympathy spontaneously, an unsearched-for phrase found by chance on the page, while the writer with a search engine goes out looking for a hand that might have reached out and grabbed their material by thinking of it first. The writer–reader relationship is anything but symmetrical, and a writer with a new sentence in their head, an idea 'thought special and particular' to them, is unlikely to feel in pleasant communion with another writer who voices it first. Although a Google search has boundless practical benefits for any writer, it can be bad for morale, swiftly disabusing them of their own particularity. A crisis in authorial originality is perhaps the profession's longest-standing fear, but the merciless way in which Google wins a game of snap has intensified the feeling that everything has already been said.

Most people could bear it if the problem of cyber-originality was limited to the writerly professions. Unfortunately, the digital age demands that all of us to some extent be writers, if to be a writer means having one's words read by scores, if not hundreds or thousands, of people. Our sentences, like never before, are the stuff of public consumption and as such the difficulties of good craft are leaving the garret and becoming part of our everyday troubles. You can see the cracking-up of some of those sentences as they're forced out into open ground, such as when a birthday goes monumental on Facebook. For some, it's an unpardonable failure of style simply to add another brick to the diligent spire of 'Happy Birthday!'s, which, while individually cheerful, adopt in unison an unexpectedly po-faced tone. The relentless, homogenised chant of social media well-wishing at times comes to seem almost taunting. Can the fiftieth Happy Birthday! hold on to its sincerity, or do its long string of predecessors give it an unintended sarcasm? The pressure we feel when writing these salutations is the pressure of the artist. How do we make it new?

With there being limited synonymic room for manoeuvre, Happy Birthday! itself erodes under these pressures of individuality into nonsensical variations – Happy Happy! Happy Buffday! Appy Burpday! Hoppy Bidet! Hippo Birdie! – which grow ever more Beckettian. It is only a matter of time until the sentiment declines into 'Half-Price Brisket!' at which point we may be able to call the whole thing off. There was far less opportunity for dismay in the old-fashioned collection of greeting cards arranged along the mantelpiece. When Dorothy Parker's friend Mary gave birth, Parker famously sent her a telegram saying 'We all knew you had it in you.' In the old days, even if your heart could only muster platitudes, at least they weren't broadcast alongside the Dorothy Parkers. Now it's as if magic dust has been sprinkled over the mantel, and all the cards are swivelling on their edges to get a look inside one another. This sort of self-consciousness is, on balance, an unhelpful development in the business of observing generic life events.

The slinging of our poor platitudes into the arena may eventually make constant stylists of us all, but there equally may arise a resistance movement. We may see defiance. Modes of communication may occur that, when it comes to originality, belligerently throw in the towel. At the opposite pole to the Happy Birthday! decadence is the nihilism of the Yo app, which impishly made itself known on April Fool's Day. This app began as a minor challenge between a developer and his boss, who wanted to be able to summon his secretary with the click of a button. Thanks to over a million dollars of seed money, this monosyllabic app now has all sorts of meanings. Friends can send Yos to each other as either text or audio alerts, giving up any pretence of Wildean flair. And if you sign up to services that use it, the Yos you receive on your devices will bear different implications depending on their source. A Yo may mean that your team has scored a goal, that a theatre has last-minute tickets, or that your Singapore spicy noodles are ready for pick-up. The Yo is an early example of style-resistant discourse, its wilfully generic exterior containing an infinite variety of meaning. In this way, Yo offers an escape from the demands of literary style. It is a concession to our age's crisis of idiosyncrasy, and while it may soon be a forgotten fad, its spirit of revolt will continue as we head deeper into the digital. For Yo never asks 'what next?'; it rubs its back rapturously against its own dead end. This is a noble approach to the problem of continual novelty in a world composed of finite aesthetic possibilities.

The problem of where to go next carries a sort of ecological foreboding. The twentieth-century Irish painter Patrick Collins conceived artistic originality in terms of landscape and habitat. He thought about the Celtic imagination as 'an untilled field' and spoke of the 'unmuddied stream of art in Ireland'. The particular character of light falling on a drumlin in a certain corner of Ireland, in other words, had not already been annexed by Monet or Cézanne. This manner of framing the problem suggests both Collins's relief at the virginity of his subject and the artist's inevitable territoriality. Newness in art is in this way a matter of finding arable land that has not been overworked to the point of producing nothing but rows of wilted clichés. But always cast onto the image of the unmuddied stream is the shadow of the artist setting up an easel on the bank. If we're to accept this analogy, then we also have to accept that all creativity leaves mud in its wake.

Evaporation Chic

Avram Davidson, an American writer whose work defies easy categorisation, also conceived cultural progress in ecological terms.

His 1961 short story 'The Sources of the Nile' is about a New York journalist who discovers that all the hot new things in American fashion are originating from a suburban, post-war family living in the Bronx. He visits this family, the Bensons, and witnesses their effortless hipness: the father has a 'soup-bowl' haircut; in a flourish the mother makes a chic turban out of a West Indian madras; the daughter, whose dress looks to be out of a Victorian children's book, is lacquering stars onto her toenails; the son is bare-chested and in denim shorts, with 'cut-off sneakers' on his feet. 'With absolute unconsciousness and with absolute accuracy,' Davidson writes, 'they were able to predict future trends in fashion.'

The Bensons sound great. In their genesis story, the fall comes when the writer who visits them gets commissioned to document their style. The namer, once again, can't leave well enough alone. After securing an advance based on his preliminary findings of soup-bowl haircuts et al., he returns to the Bronx for more, only to find that 'The house was empty. It was not only empty of people, it was empty of everything. The wallpaper had been left, but that was all.' This disappearing act is one solution to the problem of overworked land. When the world is chock full of things and open ground is scarce, there is always space for no thing. The writer in the story wonders where the Bensons have gone. Are they 'so near to hand that far-sighted vision must needs forever miss them? In deepest Brooklyn, perhaps . . . or in fathomless Queens, red brick and yellow brick, world without end.'

Given Davidson's current obscurity, it would be surprising if Alice Goddard, co-founder of the magazine *Hot and Cool*, intentionally performed a riff on Davidson's forgotten story fifty years after its publication. And yet, the echo resounds in her photoessay from the spring 2013 issue. Using Google Maps Street View, Goddard zoomed in like an omniscient narrator on a small American town and discovered her own source of the Nile. Google's roving camera cars had captured, as accidental detail, the people of the town going about their business, and Goddard was inspired not by the vision but by the normalcy of their style. The photoessay is an exhibit of everyday suburban looks: an elderly person, face discreetly blurred, hikes up his or her trousers to the armpits, two columns of grey material sagging between a sky-blue fanny pack and white Velcro running shoes. There's an array of billowy, nondescript shirts tucked into baggy jeans, a man in an oversized sports jersey, another in a Nike t-shirt and track-suit bottoms, the proportions and fit being so unadventurous that, apparently for Goddard, they had swung round to cutting-edge.

Here then is what the digital age did to the Bensons, retrieving them in that strange space of Google Street View, vision that is both far-sighted and near-at-hand. In these two cases we have the unwitting suburban style gurus, but there is a crucial, telling difference between the 1960s and now: in Davidson's story, the blessed people are unselfconscious and visionary, whereas for Goddard finding unselfconsciousness in a digital age was in itself more than visionary enough. While originality may have been elusive, at least for Davidson it was still out there, somewhere.

Goddard enjoyed this Google street-shoot-by-stealth because it robbed its subjects of knowingness. In New York, Bill Cunning-ham is a famous chronicler of street style, for whom, according to *Vogue* editor Anna Wintour, 'everyone gets dressed'. In Goddard's Anytown, USA, people don't get dressed for Bill; they get dressed for the shops. But whereas the Bensons' uncanny talent drives fashion into the future, Goddard's townspeople predict the present. Their total, natural embodiment of the mainstream is their radicalism; they might be described as *garde*. In 'The Sources of the Nile', the joke is that a parochial environment has produced a hub of progress, while in the *Hot and Cool* editorial the potential joke is that the categorically unfashionable are the new fashionistas. Fitting in is the new standing out. Some label Davidson's story as science fiction, and it wouldn't be unreasonable to suspect Goddard of satire. But her showcasing of the mundane is only one instance of a larger trend towards disappearance, which the American cultural forecasters K-Hole named, to much acrimony, Normcore, and indeed commentators have feared that Normcore too is one colossal joke.

Normcore was born in the K-Hole trend forecast, 'Youth Mode: A Report on Freedom', which is sensitive to the ongoing fear of finite creativity: 'There's a limited amount of difference in the world . . . The anxiety that there is no new terrain is always a catalyst for change.' Here again is the problem framed in terms of a shrinking landmass. One hears too an echo of John Barth's 1967 essay 'The Literature of Exhaustion', in the notion that a chronic mood of fatigue, whether valid or not, is ironically what motivates cultural replenishment. For K-Hole, the change that has been catalysed is an embracing of collective versus individual style. In a digital culture of global image-sharing, it is ultimately one's image that is shared. It becomes as tricky to believe in the *personality* of style as it is to think that you're the only one imagining Thelma and Louise in a Google Car.

Normcore's spirit might be described as the 'When in Rome' school of style and behaviour. Instead of 'making statements', K-Hole proposed, people would increasingly adopt the sartorial standards of whichever community or subculture with which they were, at that moment, engaging. Khakis for the hiking group by day, leather in the biker club by night. Notice that the old ideal of the shape-shifting cyber-self, having been thwarted by the Medusa eyes of Web 2.0, is transmuted in the Normcore vision of a real-world body moving fluidly between multiple identities. For subscribers to Normcore it is in real life, not online, that the fixed body is disappearing. It is as though the dream of John Perry Barlow's liberated cyberspace is happening not in the ether but here on earth. Normcore celebrates, as did Barlow in his 'Declaration', a world of bodies that can cross borders and boundaries. The manifesto for fluidity is reversed. With digital culture solidifying our actions, freezing verbs into nouns, with every online thought gaining the weight of endorsement, with movement cramped in an environment choked with 'things', the idea of Normcore is that the possibility of evaporation has migrated from Barlow's cyberspace and onto our real bodies.

In a world more socially connected than ever before, it is hard for niche tastes to maintain their exclusivity. This is the problem of hive-cool, where stylistic trends soon lose their cachet from being transmitted so effortlessly to everyone else. Individuality, such as it ever was, is inevitably a threatened ideal in a digital age, whose spirit of communality challenges the independence necessary for Indie-styles to occur. In *All That Is Solid*, Berman, quoting Nietzsche's ideas on individuality, writes:

> Modern man's sense of himself and his history 'really amounts to an instinct for everything, a taste and tongue for everything'. So many roads open up from this point. How are modern men and women to find the resources to cope with this 'everything'? Nietzsche notes that there are plenty of 'Little Jack Horners'

around whose solution to the chaos of modern life is to try not to live at all: for them, '"Become mediocre" is the only morality that makes sense.'

If we remove the judgement from our idea of mediocrity, 'Become mediocre' is the morality of Normcore. And, indeed, if ever a creature had 'a taste or tongue for everything' it is the 4D human. We are trained by our digital environments to sense every *thing*, and so Normcore is understandable as a reaction against this solidification of life, advocating a sublimated state where there is no *thing* left to be named. It has been described, in fashion terms, as 'the look of nothing', and one of K-Hole's tag lines for it is 'The New World Order of Blankness'. Am I the only one thinking now of the bleached-out final frame of *Thelma & Louise*? For the love of God, say yes.

It became instantly fashionable to despise Normcore, whatever it might mean, but this backlash is one of its most significant qualities. The general disgust reveals a malaise whose source is the noun invasion. 'Dressing normal isn't a thing,' said an exasperated Daniel Spagnoli, who designed a joke 'removal' app that replaces each mention of Normcore in a user's newsfeed with a coloured bar. Around the world, fashion blogs became sheepish about reporting this trend. 'We covered it. Sorry', wrote the *Daily Dot*, while Australian columnist Anna Byrne announced Normcore with 'apologetic quotation marks'. The amount of irony in the cultural blood had, it seemed, reached intolerable levels. For some, irony confined itself to a depressing and insulting evolution of hipsterism: people in Williamsburg, Brooklyn, for example, dressing for lolz like Midwestern tourists, shoulder-padding 'the everyday' with their own arch quotation marks. But this visceral reaction is more a symptom of the broader epidemic of naming, which produces here, via the self-consciousness of the Normcore term itself, the irony of conspicuous camouflage. An attempt to sidestep labelling, to become in certain ways invisible, couldn't evade the taxonomist's flicking fingers. In other words, one possible response to cultural solidification – the evaporation of our distinguishing, nameable features, our species markings – became itself a solidified pose.

Normcore was reviled, not because people didn't yearn for blissful evaporation, but because they saw it as a grotesque incarnation of the very solidification it proposed to boil off. It was, perversely, both the child and the supposed enemy of our desire to produce *things*. The attempt to make a thing of *no thing* was an abomination in the procreative naming game. A strange morbidity surrounded it, a feeling that this thing was dead as soon as it came alive. All the tag lines about freedom and blankness amounted to one large toe-tag. In a 1916 essay, Walter Benjamin wrote that over-naming 'is the deepest linguistic reason for all melancholy'. His argument was a theological one, referring to how our human babble of languages distances us from divine reality, but the loathing of Normcore was in part a hitting out against a similar melancholia. Normcore became the scapegoat for a culture of over-naming, a focal point for our endemic breathlessness in a world that, at any one moment, is choked with nouns.

Despite Spagnoli's protests, Normcore, by being echoed through cyberspace, had resoundingly materialised, into a monstrous, dead-alive thing. By September 2014 the *LA Times* could say with confidence that 'Normcore became a thing this year.' It can be hard to dissolve such things, once they have appeared. Spagnoli's app, like many attempts at effacement, only serves to highlight that which it seeks to efface; each scratching out draws the eye to the telling gap. Taken simply as a fashion trend, Normcore seems to have been eclipsed, at the very point of it being widely recognised, by its opposite. And yet, it persists, haunting its own negation. Describing fashion's autumn/winter 2014 in the *Guardian*, Morwenna Ferrier christens the new trend of LOL-core, a turn to vibrant prints, dresses stitched with absurdist designs, denim made gaudily piebald with iron-on patches. Again there is the apology: 'Maybe', Ferrier writes, 'it's a reaction to Normcore (yes, we said it).' One commenter on LOL-core posted, at the blue-lit hour of 2.47 a.m.: 'In my opinion, any trend with the suffix "-core" in it deserves immediate contempt.' Ferrier thinks of LOL-core as 'things on things', though she admits that, since the designer J. W. Anderson created a line for Topshop, back in the mists of 2012, '"things on things" has been a thing in our wardrobes . . . but the newest things to appear on a thing are faces'. In this sense, Normcore wasn't a rebellion against a culture of thing-ification – it was a bridge towards its hypertrophic opposite: things everywhere, things upon things upon things. And now, instead of a retreat into the undifferentiated mainstream, the things are stamped with faces.

You might be thinking that there is nothing really solid about these *things* at all, if Normcore lapses so easily into LOL-core. Surely those buckets of ice are long melted. Aren't we still in the maelstrom, where everything is continually boiling off? Who remembers trending hashtags from two months ago? When Twitter timelines fly past, doesn't it feel as though everything is indeed turning to air, that nothing is substantial enough to grasp? Marx and Engels claim that in bourgeois capitalism 'All fixed, fast-frozen relations, with their train of ancient and venerable prejudices and opinions, are swept away, all new-formed ones become antiquated before they can ossify.' Isn't this instant anti-quation the very definition of an internet meme, those temporary in-jokes shared online? In an environment concerned with the question of novelty, internet memes flourish because the velocity of their mayfly lives makes the difference between inspiration and expiration almost unnoticeable. Being new isn't the point of them because these little ephemeroptera are old almost as soon as they appear. They thrive on the same elements that produce cliché – overexposure, repetition – and so escape worries over being retro-grade by refusing to fight for absolute originality. In this sense they are perfectly adapted to a milieu of exhaustion; they only come into being at all by exhausting themselves. Once exhausted, internet memes are culturally biodegradable, dissolving into the air as quickly as they appeared. That is, of course, only to the extent that any digitised creature is able to dissolve.

But nineteenth-century capitalism packed the world full of ossi-fied things too. It produced plenty of new-formed obsessions, as real and durable as stone. And the same is true today. For while things certainly come and go from our timelines, from our memories, what we seem to be stuck with now is a constant aware-ness of their production. In our digitised habitat the instinct to categorise, to name, is cementing. We always have one foot in the factory. Any one moment is thus surrounded in 'things', packed inside an ever-materialising present. Despite all the great, hilarious, kind, irreverent, moral stuff that digital culture makes, it is hard at times not to feel the weight of it all. This is not simply a ques-tion of the digital trace, but the trace that is left in us. We feel the impact of digitisation's continual solidifying of the moment: the way it tempts us to plasticise the present in pictures and phrases, which then, hopefully, grow heavy with collective response. This ongoing way of making something out of the fleeting moment means that such moments never quite biodegrade as they otherwise might have. Even if we never see them again, and though we may not remember what was 'a thing' last season, we somehow remember the space they take up. With each passing year, do we not sense this unseen accumulation of solidity, swirling undigested as in some remote oceanic gyre? When digital life is overwhelming, we might be reminded of those beaches halfway round the world, where a handful of sand contains a beautiful rainbow of smooth plastic pellets.

Exceptionally Normal

Sometimes it takes a ludicrous advertising slogan to articulate a cultural anxiety. In the case of Gap's 2014 'Dress Normal' campaign, the anxiety stems from the problem of self-expression in a digital climate where only 'things' really count, and all such things have, by definition, been done many times before. An inescapable truth of a hive society, from the egotist's point of view, is that everything is done to death.

Gap's command to 'Dress Normal' is as off-putting as a lover stroking your hair and whispering, 'Relax,' and as alarming as being small and watching a parent slump in an armchair and say, 'Eat what you want.' It appears to signal its own creative exhaustion. However, we shouldn't be fooled by false fatigues. Gap, with its corporate eyes wide open, is trying to grapple with a genuine cultural weariness over the difficulty of being an individual in a networked world. It, like us, is having trouble reconciling the excitements of connected, intertwined living with the long-standing pressure to be individuals. As a result, what appears is this nonsensical slogan, which tries simultaneously to celebrate hive life while refusing to abandon that age-old appeal that the commodity makes to the customer's exceptionality.

The campaign includes a series of noirish adverts directed by David Fincher, in which clutches of abnormally beautiful people engage in urbane, saucy scenarios. At the end of each we are given subheadings to the Dress Normal banner. Some have a puritanical flavour: 'Let your actions speak louder than your clothes' or 'Dress like no one's watching'. For those who like all their bases covered, there is: 'The uniform of rebellion . . . and conformity'.

In Benjamin's essay on naming, the natural world is sorrowfully, deliberately mute, being over-named by 'man' and all his languages. Naming, for Benjamin, brings mournfulness to the thing named, since human names always misrepresent the thing and drive it further from the *real*, divine name. Sorrow gathers in the space between the thing and its mortal names. Everyone is as dumb as animals in the Gap ads, but their muteness is full of jubilation, sponsored by Eros. They are after all talking with their actions not with their labels, verbs on the counter-attack. And why shouldn't they be happy, having shrugged off society's sartorial categories? They have been, as Benjamin would put it, 'fertilised', released from their melancholy, and so they dance and kiss and pull off their trousers. They are showing us that communication occurs with Benjamin's 'mute magic of nature'.

Benjamin would certainly be interested to hear that Gap's chief marketing officer, Seth Farbman, believes that 'my normal is different from your normal'. In this set-up, the world isn't over-named but under-named, with 'normal' standing for a limitless number of different things. If we take the Dress Normal universe at face value, the Namer seems to have left the building and naming has ground to a halt: labels are finished when rebellion looks the same as conformity, and when normal is literally meaningless.

According to Farbman the way to 'your most authentic self' isn't a matter of being called by the perfect name but 'finding those perfect items – a pair of jeans, a t-shirt'. Note the sound of an old idea taking shape inside this revelry. That ultimate label of the 'self' is back already, the 'I' that is the biggest name of all, which in its authenticity is different from all others. Finding the way to your own personalised, subjective 'normal' is individualism by another name. Thus the strange nonsense of Dress Normal, like a versatile coat, can be turned inside out. When everyone's normality is unique to them, the normal is necessarily the exceptional. The billboards could equally read: Dress Exceptional. Here is the irony of Gap's command. In one ad, the words 'A simple jacket for you to complicate' run over an image of Zosia Mamet lying on a white sheet, inert as a knocked-down child in road-safety posters. But despite this pose of corpse-like indifference, the slogan hints at the revival of an individuated, egotistical will, gathering strength, consolidating itself within a costume of explicitly ironic mediocrity.

Nietzsche, in his comparable suggestion to 'Become mediocre', claims that mediocrity has something to teach the exceptional. He sees mediocrity, in the context of Christian democracies, as a position that the exceptional must learn to adopt in order to wield power. Something akin to a millionaire prime minister telling the people that, when it comes to poverty, 'We're all in this together.' The problem with this masquerade, however, is that mediocrity, for Nietzsche, 'can never admit what it is and what it wants. It has to speak of measure and dignity and love of neighbour.' And so the exceptional, dressed as the mediocre, 'will have trouble *to conceal the irony*!' The Gap campaign has trumped this problem by not bothering to conceal anything. Its irony isn't hidden, but worn on its sleeves. The ads have therefore supplanted the Nietzschean vision of discreet mediocrity with one that *does* admit what it is and what it wants: to be normal in its own special kind of way. Owing to the digital age's great ambivalence over the possibility of a 'networked individual', the irony has become too glaring to conceal. And, as we have seen, this ambivalence is plugged into the larger irony that the dematerialising effects of digital technologies run concurrent with a solidification of experience and our articulations thereof. The digital, in one stroke, manages to sublimate the real and petrify the ethereal. As a result, we feel our lives evaporating and solidifying in the same breath, and as when something very hot meets something very cold, we feel cracks beginning to form.

Or perhaps that's just me.

4

Keeping and Killing Time

'Now *all* the petrol has stopped and we are immobilised, at least immobilised until we get new ideas about time.' This was how the author Elizabeth Bowen described wartime life in Ireland to Virginia Woolf, in a letter from 1941. Bowen explored some of these new ideas in her London war fiction, which is full of stopped clocks and allusions to timelessness, the petrifaction of civilian life in a bombed city. Across the literary Channel, Jean-Paul Sartre's war trilogy, *The Paths to Freedom*, is, like Bowen's Blitz work, in part a study on how time itself becomes a casualty of war. In one scene Sartre describes German troops ordering a division of captured French soldiers to adjust their watches to their captors' hour, setting them ticking to 'true conquerors' time, the same time as ticked away in Danzig and Berlin'. Historically power has been wielded through the pendulum, and revolutionary change has been keenly felt through murmurs in the tick and the tock of one's inner life. King Pompilius adjusted the haywire calendar of Romulus, which had only ten months and no fidelity to season, by adding January and February. Centuries later, the Roman Senate renamed the erstwhile fifth and sixth months of the Romulan calendar to honour Julius Caesar and Augustus, thus sparing them the derangement still suffered today by those once-diligent months September–December. For twelve years, French Revolutionaries claimed time for the Republic with their own calendar of pastorally themed months, such as misty Brumaire and blooming Floréal.

The digital revolution likewise inspired a raid on the temporal status quo. In 1998, the Swatch company launched its ill-fated 'Internet Time', a decimalised system in which a day consists of a thousand beats. In Swatch Time, the company's Swiss home of Biel usurps Greenwich as the meridian marker, exchanging GMT for BMT. This is a purely ceremonial conceit, however, since in this system watches are globally synchronised to eradicate time zones. A main selling point of BMT was that it would make coordinating meetings in a networked world more efficient. This ethos severs time from space, giving dawn in London the same hour as dusk in Auckland, and binding every place on earth to the cycle of the same pallid blue sun. As it turns out, we didn't have the stomach to abandon the old minutes and hours for beats, and the Swatch Time setting that persists on some networked devices is the vestige of a botched coup. Although this particular campaign was a failure, digitisation is nonetheless demanding that we find our own 'new ideas about time'. For as the digital's prodigious memory allows our personal histories to be more retrievable, if not more replicable, we are finding in the civic sphere a move towards remembrance that shadows the capacity of the network to retain the past. But while time is not lost in the ways it used to be, the tendency of digital technologies to incubate and circulate a doomsday mood is much.' In the pause before Karim announces the really long thing that elephants possess, I expected him to wrong-foot me not with trunks but with memory. The powerful recall of elephants is both myth and scientific fact. Aged elephant matriarchs, in particular, are able to recognise long-unseen faces, as well as reacting to present dangers by invoking successful tactics from the past. It's therefore apt that elephants were present at the birth of YouTube, since the website has become a vast repository of memories.

I should say at the outset that YouTube isn't only a gateway to retrospection. It skims the surface of the moment, catching the very latest of our shared interests and concerns: street footage of political unrest around the world, insurgence, injustice, potent speeches, confessions, showbiz gaffes, civilian charm and civilian embarrassment. It is also instructional, showing you how to master a bow tie or rewire a light socket. I've known doctors who will go on YouTube while a patient is being prepped for surgery, just to get a quick refresher course on certain procedures; although that is something I try to forget. But another of YouTube's executive roles is archivist of celluloid history. With the steady industry of uploading vintage television, films and even adverts – who *are* all these kindly gnomes, mining the old times in this way? – the past is making a comeback. There is some fluctuation, as the copyright police enforce their purges. Before the big bosses fully understood the scale of YouTube, you could watch entire television series that were still bringing in hefty syndication fees elsewhere. However, the exponential expansion of YouTube's catalogue has overtaken such clampdowns in terms of sheer bulk. It even thinks of itself as time's warehouse. As I write, YouTube's own statistics page estimates that a hundred hours of video are uploaded per minute. At that rate, over sixteen years of content are added to the archive each day. Its Content ID system, which cross-references uploads with a database of copy-righted material, 'scans over 400 years of video every day' to check for infringements. While you won't get far looking for unsullied bootleg *Friends*, older, long-lost pals are gathering there to stoke your nostalgia. A 1980s childhood, for instance, is currently being resurrected on YouTube.

The resurrection is occurring piecemeal. I remember when the world was young and YouTube was sparsely furnished. In the early days, around 2006 and 2007, I would check it now and then, to no avail, for *The Box of Delights*, a BBC Christmas miniseries I'd seen just once at the age of about five or six. Only two scraps of detail remained in my mind. The first: a real old man on a real pony trotting inside a painting, the duo becoming painted themselves and disappearing around a mountain path; the second: an old woman scolding someone. 'Your wolves are frightening my unicorns!' For years, whenever these two memories would come to me, one after the other, the thick fumes smoking off them would give, in comparison, an exquisite emptiness to the present. Theirs was the heyday of my capacity for magic, no doubt enlarged by the time of year, when my secular schoolboy head was restocked with stars and donkeys and little babies up in the bright sky. Not being able to remember anything of the room I was in when I watched *The Box of Delights*, I tended to absorb these two rootless memories into a more vivid atmosphere from a few years later: a Sunday evening in the acreage of my favourite childhood living room, *The Chronicles of Narnia* piping away on the television, my mother nearby, reupholstering a rocking chair. To borrow from Amazon, 'Customers who think of *The Box of Delights* also think of *The Chronicles of Narnia*.' But then, one day recently, my call down the well was answered with more than an echo. A YouTube search replied with six uncanny little stills, my trained eyes knowing at once from their ample durations that they represented intact fossils.

The opening scenes of Episode One used the familiar tropes from this English Yuletide genre: a steam train chugging through a faded winter's landscape, boarding-school homecomings, boys' acting voices all posh and reedy, plucked harp-strings denoting the supernatural, trumpets denoting supernatural majesty. How quickly tedium set in. I was impatient, not for what I had forgotten, but for what I could half remember. I began to skip the story along, looking for the old man and the painting. When I found him I did feel a frisson of recognition, but the overwhelming sensation was of estrangement from that potent atmosphere of my memory. Still hopeful, I went chasing after the old woman, tracking her from episode to episode across that flip book of frames that one riffles above the time bar. Everyone knows by now that you can never go back, so I'm not sure why I needed to live out this anticlimax. YouTube is especially insistent on the other-country quality of the past because it makes you view the past literally through the prism of the present. Loading up Episode Five, its title 'Beware of Yesterday' being on-trend, I continued my hot pursuit of the elegant old sorceress, with a semi-transparent banner for 'Accountants & Taxation' obscuring the lower strip of the screen. My nostalgia thus imprinted with the Google-search incarnation of my current woes, I discovered eventually that I had got the line wrong anyway: 'Keep your lions away from my unicorns,' the old woman warns Herne the Hunter as they course through the sky on sleighs drawn by their respective takes on flying reindeer.

In the film version of *Shirley Valentine*, Shirley fulfils a fantasy of sitting at a table by the shore of a Greek island, drinking wine and watching the sunset. In the daydream she knew exactly how it would be. However, she admits, 'Now I'm here it doesn't feel a bit like that. I don't feel at all lovely and serene, I feel pretty daft, actually. And awfully, awfully old.' The future, when it comes, can make fools of us in this way, but equally so can the past.

I'm able now to think of a cartoon that impressed me as a child and within seconds be fairly sure of finding at least snippets of it on YouTube, or perhaps dubbed into other languages, so that my dear Super Ted has apparently learnt Swedish in the time we've been apart, and the Thundercats now roar in Italian. As I learnt from *The Box of Delights*, a snippet is often enough; a blast is all we really need from the past. For all its power, nostalgia is strangely lazy, and memory lane runs on a downhill slope. Had I wanted to, I could have tracked down my after-school companions, ordering DVDs online and assembling a retrospective, but those lost feelings were never quite worth the bother. Now, however, these fragments of the past are for the first time on tap, not stored away in boxes. Not since storybooks dominated childhood life have we been able to challenge the nostalgia for our earliest days by satisfying it so relentlessly. Although for decades we've had the ability to rove through the old times through song, in the first years we don't develop the self-conscious soundtrack to our lives that encases teenage terror and ecstasy in amber. Opening theme sequences were among the first messages that my generation sent out to greet our future selves with a pang, coming at us as they did, day in, day out, as we rolled around on the carpet. What's more, YouTube can move almost as quickly as our leaps from memory to memory, so that we can curate an external exhibition of our reminiscences.

About a century ago, Marcel Proust thought it necessary to write five pages on the experience of biting into a piece of cake, and this moment has come to symbolise a certain type of remembrance that results unexpectedly from a sensory trigger. The narrator of *In Search of Lost Time*, also named Marcel, is given tea and the now-notorious scalloped French cake called a madeleine, the taste of which summons up a whirl of images from his childhood. Marcel describes how

> in that moment all the flowers in our garden and in M. Swann's park, and the water-lilies on the Vivonne and the good folk of the village and their little dwellings and the parish church and the whole of Combray and its surroundings, taking shape and solidity, sprang into being.

With YouTube's enveloping of the past, it's as though I suddenly have a YO! Sushi-style conveyor belt of evocative confections running around the walls of my home, with slices of Battenberg, Mr Kipling's Six Rich Chocolate Slices, a melting Viennetta, my nursery school's apple crumble, Penguin bars and the rest, all gliding past for my sampling. However, Proust's point about the madeleine was that its effects were unplanned; the taste of it provoked a *mémoire involontaire*. So many of our past-blasts now are not only voluntary but wilful. Marcel realises that merely seeing the madeleine had not provoked his memory because although he hadn't eaten one in years, he had seen them regularly, 'on the trays in pastry-cooks' windows'. And so, 'their image had dissociated itself from those Combray days to take its place among others more recent'.

With this ever-present quality of our digitised pasts, do we leave ourselves the space in which time can be lost and then surprisingly recovered? We remember those shady dolls tottering about in stop-motion partly because we assumed that they were gone for ever. Certainly for people much younger than me, the cultural past has never fully left, artefacts from the old days never totally irrevocable.

In Philip Roth's *American Pastoral*, the narrator Nathaniel Zuckerman describes the uncanniness of attending a forty-five-year high-school reunion in 1995, just as the early web was being spun. Classmates laugh and scream as they try to dig each other's young faces out of sixty-something landslides. Nathaniel begins to view the whole thing as a joke, 'as though "1995" were merely the futuristic theme of a senior prom that we'd all come to in humorous papier-mâché masks of ourselves as we might look at the close of the twentieth century'. Thanks to Facebook, those classmates are now coming with us, and we'll watch them age in instalments. Reunions of the future will lack some of the explosiveness of Roth's ball, where 'time had been invented for the mystification of no one but us'. Instead they may be more like animated and hopefully more tender versions of the subdued, retro mingling that goes on in social-media newsfeeds. In the scramble of Facebook migration, many old faces from my school and university days that would once have been lost in time have stowed aboard. True to life, we never interact, but I keep them there almost superstitiously, and they pop up now and then in the trench of my time-wasting to advertise the latest increment of their aging. Even my old friend *The Box of Delights*, I notice, has a Facebook page.

YouTube's memory for your private passions is another way in which a version of yourself becomes solidified online. Lipstick on your sidebar told a tale on you. I would blush if someone who didn't know me well logged onto my YouTube account and saw the rogue's gallery of suggested videos, where Bea Arthur huddles next to Lion-O, both of them cold-shouldering Dogtanian. But it is not only the private past that is being regained; we are also in a period where the scope of public remembrance is being redrawn.

the most of digital memory. Even if we have yet to reach Lanier's world of micropayments, we know by now that digitisation is making the past accessible and retrievable in unprecedented ways. We're aware of the potential indelibility of our own digital traces, how this fourth dimension tattoos us as we move through it, and indeed how we mark it in turn, how its surfaces are more impressed with handprints than the Hollywood Walk of Fame. Every pathway in the network is a miniature memory lane, and as a result we see ongoing legal debates about 'the right to be forgotten', which in the case of online life includes the right to make Google uncouple you from your past. This right states that we aren't obliged to be permanently associated with everything that a Google search of our name could dredge up, such as compromising or distressing photographs. If digital life remembers more than we would like, then at least, according to Lanier, we should be compensated. But what are the implications of trading in a currency that remembers everything that ever happened to it – every journey it has taken around the world, every palm, every pocket? An image of money arises that is as well documented as we are, with every handshake instagrammed, every transaction memorialised in a status update: 'Looking forward to making these much-needed luxury flats happen! – with *Smiling Corporate Sponsor* at *Unprofitable Hospital.*'

In the early months of 2011, two events, which differed greatly in terms of global significance, both gave a preview of what life might be like in a digital world where money comes with a history. The first event was the disgrace of fashion designer John Galliano; the second was the West's reaction to the Libyan War that would ultimately depose Muammar Gaddafi. Setting these two cases together is ridiculous in all ways but one, for they both uncovered the shadow-side of our connectivity, and how the insistence of the past in our digitised present will increasingly demand a new sort of morality.

In February 2011, a video appeared of John Galliano sitting outside La Perle cafe in Paris, in an obvious state of addicted unwellness. From experience I know that La Perle is a place where it helps to have all your faculties. I once met a young man there who, when Yves St Laurent died, was so upset that he skipped work by saying his father had just had a heart attack. People take fashion seriously at La Perle, which perhaps gives it its air of impending violence. Colonising the corner of rue Vieille du Temple and rue de la Perle in the Marais, it consists of two lengths of pavement supplied with drinks from a small bar area stripped of all possible comforts, and grows throughout the evening into a fuming billow of scarves and immaculate lips and biker jackets. The fear in a place like that of being slighted by someone cooler than you reached a deranged apotheosis when Galliano, holding drunken, slurring court, professed his love for Hitler and told some off-screen interlocutors that 'people like you would be dead today'.

The scene, which happened in late 2010, was not allowed simply to occur and then lapse into the wake of the past. The digital trace, like a mischievous god, was present. Someone was filming Galliano, and the video went viral. Because of this footage and another similar incident that was reported to the police in February 2011, Galliano was fired as head designer at Christian Dior, and Natalie Portman, at that time the face of the perfume Miss Dior Chérie, said in a press release: 'I will not be associated with Mr Galliano in any way.' This is in one sense a strange thing for Portman to say, since disassociation was always the point of her contract with Dior. She was employed to be associated with little else but glamour, and certainly not with anything that had preceded the photographer flashing a camera in her face: sales reports, profit projections, target markets, campaign strategies, all those un-chic, self-interested commercial processes. Part of Portman's job was to obscure this history of the product. Arguably the luxury industry's most hackneyed adjective is 'timeless', and in one of the posters for the eau de toilette Portman strikes an ahistorical pose of feminine allure. She folds her naked breasts in her arms and regards a world of consumers over her shoulder with that look that is the only look available to her in such a context: lips three millimetres apart and irises slid to one side so that the eyes become two semi-eclipsed moons. It is the look of *Girl with a Pearl Earring*, but with the added ocular intensity of someone desperate to sell you something. But after Galliano's breakdown, time could no longer be forgotten. The spell was broken. Suddenly, this timelessness that Portman was hawking was undone by a particularly grim history of imbalance, addiction and dependency.

Two years later, in his first television interview since the scandal, Galliano told American journalist Charlie Rose that he couldn't remember the scene outside La Perle. He said that in those days he was a 'blackout drinker', whereby 'one can't transfer short-term memory to long-term memory'. So, while for Portman, as well as all those entranced by the timeless glamour of Dior, Galliano's tirade was a violent moment of remembering, for Galliano it was the opposite. And yet, ironically, he also figures the scene as a sort of rupture through which the ugliness of his own past comes rushing. By the time of his interview with Rose, his rehabilitation to that point had led him to surmise that the Hitler speech was in part a resurfacing of 'frustrations from childhood . . . being persecuted, bullied, called all sorts of names'. In the oblivion of his drunkenness, he believes, an old self-defence mechanism was triggered, and it was as though he was speaking words from the past. Growing up in a diverse population of newly arrived immigrants in South London, Galliano says, 'I would have heard those kind of taunts.'

Besides the depressing personal circumstances of Galliano's outburst, this scene outside La Perle can also be viewed as a pantomime account of what can happen when the history of financial relationships is exposed, when time comes to commerce. A relevant Marxian term here is reification, the process that makes life seem primarily organised around relationships between commodities, the things we exchange in the marketplace, rather than the people behind these commodities and their relationships to one another. Reification therefore makes us think about 'labour', 'wages', 'food', 'lodgings' as being the components of this economy, rather than me working for you, and you working for him, who used to work for her, who rents from them. As a result, the relationships between all the *people* involved in this network of exchange, including the histories of each person, are forgotten. Reification is thus the opposite of Lanier's economy of remembrance, which is based on digitisation's humane ability to personalise the commodities in the marketplace, to keep them anchored to the real people who made them possible.

To put it mildly, there is room for a certain impersonal brutality to sneak into a system in which people are forgotten. One of the ways reification happens is that such an economy of 'things' is made alluring, giving all of us involved beautiful dreams and aspirations to make us forget the often unjust relationships between people in the network of production. Isobel Armstrong, in her book *The Radical Aesthetic*, refers to 'a scene of seduction in which the fierceness of power and the brutality of capital can be disguised'. Galliano famously said, 'My role is to seduce,' and his strategy often involves the moony mystery of seducers such as Portman.

But in my pantomime version of the Dior scandal, Portman has been caught out, mid-seduction, by a bursting-forth of the ugly realities that her performance is designed to conceal. The public violence of Galliano's outburst jolts the system from its amnesia, and the world is made to remember his financial relationship with Portman, their positions as individuals in a corporate structure. Understandably, Portman can't tolerate this association. For his part, Galliano in the amateur video personifies impersonal fierceness and brutality. Recall that in his account of the event, the words he spoke had no conscious owner. In black coat, black cap and black, groomed moustache he arrives into Portman's career like Mephistopheles. He unconsciously functions in this crude spectacle as a pantomime avatar of a reified notion of capitalist ideology: fascist, indifferent and cruel. He is the obscene genius who refuses to remain offstage, the disguiser who insists on revealing himself, and whose appearance causes a breach in the fetishised image of Portman, a fissure through which history and memory comes flowing. As a cracked commodity, she can no longer maintain the coherent illusion she was paid to enact, to play her role as the self-contained object of timeless desire and aspiration.

That week in 2011, the broken home of Dior was like the Shakespearean subplot, microcosmic and thematically resonant, to a larger drama. It was a period in which time was regained and the past made vivid by violence. Within a few days of Galliano being fired from Dior, the economist Howard Davies resigned as director of the London School of Economics, over accepted donations from the Gaddafi family and his role as envoy to Libya. The Canadian singer Nelly Furtado, someone with whom Howard Davies is not normally associated, also atoned that week for her connections to Gaddafi. On Twitter she admitted to receiving a million dollars from the 'Qaddafi clan' four years previously, for a forty-five-minute private show in an Italian hotel. Furtado, along with fellow musicians Mariah Carey and Usher, offloaded the proceeds from these gigs into the absolving hands of charity. Donating a million dollars is the rock-star equivalent of clearing your history, but just as we are warned about the persistence of our digital traces, such erasure is never total.

The challenge to overthrow Gaddafi meant that the past lives of money, once so easily elided by the flow of capital, was remembered. As with Galliano, the dictator's death stimulated an undoing of reification such that the relations between people rather than between commodities were restored to public consciousness, exposing the inevitable compromises made in the accumulation of disproportionate amounts of capital. The music world's gestures of atonement certainly reflected the scale and pressures of the Libyan Civil War, but the extrapolation of their logic has interesting implications for the entire capitalist model. That is to ask, should we only sell our services to the virtuous, thereby excluding the immoral from any sort of economy of exchange? Would a waiter, if the dubious pasts of his diners were revealed, be obliged to donate his tips? What are the consequences of never forgetting? The Libyan uprising produced a rupture in the circuitry of capital that forced us to personalise and interrogate institutional systems, to examine the faces on our banknotes for signs of wrongdoing. Digital life's remembrance culture suits this interrogation. We are back again with Lanier's micropayments. In a digital information economy, where money's history must be remembered and the past is retained as an asset in the present, then there may be fees to pay as well as dividends to receive.

The idea of money having a biography is emblematic of a digital age because it emphasises our own connectivity and the ways that our paths intertwine. The fall of Gaddafi caused ripples of wickedry through the hive. It will become increasingly necessary to sever old ties in a digital world where those links can easily be traced. There has always been dirty money and political distancing, but a culture of remembrance will emphasise both. In this sense digitisation, by strapping our pasts to our backs, is demanding a new kind of virtue. It obliges us to exert in the civic sphere the same kind of mnemonic rigour as our networked machines. In the UK, Australia and the US, there has been in the last few years a wave of sexual-abuse allegations directed at high-profile male actors and entertainers, which has arisen from a complex set of conditions. The cultural and juridical shift towards believing rather than shaming victims, as well as a transferral of stigmatisation from victim to perpetrator, hasn't come from our proximity to capacious digital memory. We certainly didn't need digital technologies to expose Jimmy Savile's crimes; indeed, the approach was fittingly low-tech. Daniel Boffey of the *Guardian* notes how the head of the official investigation, Janet Smith, 'used a similar methodology to that employed during the [Harold] Shipman inquiry' fourteen years previously. Smith's team 'sent letters' to present and former BBC employees, asking them to recall relevant information. But this scandal and the other cases that have arisen more or less concurrently signify something besides a long-overdue opportunity for justice. They corroborate a broader cultural mood that feels the proximity of the past, its accessibility, a sense that it has been preserved for our moral re-evaluations. These years are thus characterised by a fervent recollection, such that lost histories are unveiled and the present becomes infused with history.

The Gaia of *Grazia*

Like the past, the news is with us more today than it was, and thus with wireless access comes a certain tariff to our peace of mind. For as long as I can remember, the news has often been a brochure of ways to die, and its instinct is to feast on both the rare and the epidemic, the two extremes of sensationalism. But now with smartphones, in particular, we are exposed to this doom media on the hoof, ensuring in our lives a steady drip-feed of apocalyptic sentiment. This morbid backdrop to daily life produces a form of unregulated consumption that is altering our ability to prioritise our fears and is encouraging, in those susceptible to such ideas, the notion that time is winding down. If the past is faring well in the fourth dimension, the future is not looking as robust.

I remember that this doomsday feeling was especially vivid on the evening of 14 April 2010, a day after the ides. It was the night that the BBC4 documentary series *Beautiful Minds* aired its episode on the radical scientist James Lovelock, who in the 1960s formulated what is now known as Gaia theory. Lovelock's general proposition is that, since the sun is thirty per cent hotter today than it was when life first appeared, our planet and its biodiverse cohort of organisms must collectively comprise a stable, self-regulating mechanism that adjusts homeostatically to such changes in conditions. After almost an hour of charming biography and interviews, the nonagenarian professor announced with grandfatherly calm that a reduction of our carbon dioxide emissions will be of little use in the climate-change project. According to Lovelock, when it comes to causes of global temperature increase, the pitter-patter of our carbon footprints is nothing compared to the damage already sustained to this calibrated Gaia mechanism. He predicts that heat which once would have been reflected back into space by miles of Arctic ice will now, with these brilliant mirrors rapidly melting, be absorbed by the surrounding waters. Our efforts to reduce carbon, therefore, will be but a drop in a scalding ocean. Furthermore, Lovelock, at ninety, felt that it is not impossible that he will live to see major adverse changes manifest, and in his view the situation is 'fearsomely bad', estimating it could eventually kill seven out of eight of us.

Two hours before this episode on Lovelock was aired, police were called to the Hoxton Chicken and Pizza shop in Hackney, where sixteen-year-old Agnes Sina-Inakoju had just been shot in the neck. Someone on a bike outside had fired a bullet through the window. I learnt about the shooting the next day, on my way to checking my email. Two days later someone posted on social media that she had died in hospital. It transpired that she was the accidental victim of a gang-related revenge attack. Online I learnt about her skills in public speaking, her teachers' high expectations for her GCSEs. As I googled her name to find out more about the story, an echo of Lovelock's grim prognosis resounded in my mind, because it struck me that there are so many simpler, swifter ways to die than the elaborate, Bond-villain scheme of melting icecaps leading to a chain reaction of starvation, disease and territorial war. Forget the four horsemen: one day you may walk into your local fast-food outlet with friends and never walk out again. As can be the ignoble way with these things, I began to feel sorry for myself. I thought about how many times I had stood at the counters of East End takeaways, fuelling up for the night bus, absurd and conspicuous from the evening's excitements. And then there were the icecaps. And it was the week of the ash cloud. My hand went to my neck; I looked up through the empty sky towards the sun.

The collision of these two stories is just one example of how, in an age of constant information, we must daily reconcile two scales of tragedy: the personal and the planetary. The digital age supplies us with a steady exposure to two infinities of horror, the universe of sorrow contained in individual loss, and the vast dread of our collective undoing. Coverage of news stories such as the Sina-Inakoju shooting is a sort of magnification, whereby the once-invisible particulars of one Londoner among millions are projected into our visual field. Sina-Inakoju's face appeared in the photograph accompanying her story, and for all the wrong reasons I was able to see the glint of her nose-stud, her cropped haircut and her smile, the happy tilt of her head that made her earring brush her collar. While Sina-Inakoju's death was a freakish combination of timing and circumstance, it was presented in main-stream news media not as an isolated tragedy but as evidence of something larger. A *Telegraph* article pointed out that she was 'the fourth teenager to be murdered in three weeks in London'. This arbitrary grouping of data (why stop at three weeks?) kindles the perversely seductive feeling that *things are getting worse*, that we are all speeding towards some calculable end. In the same year, this doomsday mood was sufficiently widespread for a Professor Spiegelhalter to hold a mirror to our paranoia by publishing a statistical study showing that London homicide crime was *not* gathering any sort of momentum.

Technologies that extend our ability to bear witness have histor-ically produced melancholy and unease. Over 350 years ago, the morose Blaise Pascal, who may have been more inclined than many to such emotions, was having a similar problem with 'new media'. Advances in lens technologies had brought two opposite extremes of scale hurtling into the human eye. Philip Fisher, in *Wonder, the*

Rainbow, and the Aesthetics of Rare Experiences, explains how, for Pascal, 'The new microscope of Leeuwenhoek and the new telescope of Galileo reawaken a theological horror of the infinities they reveal, which, instead of exhilaration and curiosity, unleash profound depression and terror.' After centuries of Enlightenment thinking, we can just about tolerate walking around in a universe that contains subatomic particles and red giants. But the digital age has produced its own problems with optics, which ultimately affect how we look at the future. The very concept of 'trending now' is a magnifying glass, and newsfeeds and timelines are bespectacled with an array of lenses, able to accommodate all sizes of terror in their frames. They zoom in on the particular, the isolated tragedy. News of untimely death travels across social media; there is always a memorial service for someone your age, a friend of a friend of a friend. At the same time they zoom out, making visible ISIS's advances, Eurozone meltdowns and shrinking icebergs.

The two stories of April 2010 illustrate this Janus-face of modern reportage, with one side pointing a telescope to the poles while the other studies the lab slides of individual human sorrow. One face beholds the Chinese industrial boom and its clouds of coal, the other peers into the specifics of personal tragedy, at Sina-Inakoju's star necklace and her stolen academic promise. If the gamut of our information sources produces an environment favourable to fear, then it is a habitat of practically unbounded scale. We daily consume the catches from a net that trawls the universe for terror, its mesh able to scoop up both H1N1 and Asteroid DD45.

This collision of scale threatens us with a loss of proportion, a precondition for a culture of panic. In such an environment, our futures are continually called into question by the many possibilities of how they may never arrive. Panic is connected to time because it breeds inside stories, and stories are temporal creatures: composed of beginnings, middles and ends, strung together by cause and effect. Through the engine of narrative, panic propels us in imagination to the end of our own stretch of time, into our own non-future. Soon after 9/11, which ushered in a new global culture of panic, Ian McEwan wrote an essay about how the last words of the victims allow us to enter imaginatively into their terror:

> you are under the bedclothes, unable to sleep, and you are crouching in the brushed-steel lavatory at the rear of the plane, whispering a final message to your loved one . . . You have very little time before some holy fool, who believes in his place in eternity, kicks in the door, slaps your head and orders you back to your seat. 23C. Here is your seat belt. There is the magazine you were reading before it all began.

McEwan sees this imagining as 'the mechanics of compassion', but it is also the mechanics of irrational fear, when you feel that any moment you could cross the fourth wall into the story of annihilation. Their disease is my disease, their memorial service soon to be my own. In this sense, panic is perhaps our most instinctive mode of literary thought, of a frantic mind quickly spinning dread stories about the monsters beneath the bed. It is the mode that turns the silhouettes of dressing gowns into midnight assassins replete with histories and motives, the agents of impending, bloody climaxes. The story's transmittable form is precisely what makes horror airborne, contagious to imagination. It was a contagion of narrative that sent Norman Bates tiptoeing into the bathrooms of the world. We substitute ourselves for the news's daily victims, or imagine ourselves bracing for the final story, the narrative counterbalance to the Big Bang. A central apocalypse myth is a defining cultural feature and, as befits a consumer society, we have a choice of competing endings to buy into. However, what makes our own doomsday gloom so peculiar is the four-dimensional twist at its palpitating heart. For on the one hand our range of digital media furnishes us with limitless paths to oblivion, and yet through the very form of this media we are immersed in a self-replicating, repetitive process without end. We are trapped in a paradoxical relationship between content and form, whereby we are endlessly being told that the end is nigh. We live at once with an acute sense of an ending and also a sense of never-ending.

Alongside a post-9/11 culture of panic, there has been the concurrent intensification of its opposite: an obsession with the mere possibility of the *event* that has come to be known as the culture of celebrity. Web 2.0 is growing up amid these two forces. Celebrity culture sustains itself as reportage stripped of content; it is the news without the story. It continually pulls us out of time by making us focus on the non-event, the non-happening, exalting in the formal processes of reportage in an endless, simple worship of occurrences. While panic culture thrives on stories about nothing more happening to unfortunate people, the cornerstone of celebrity gossip is the joy that even the most mundane happenings are still possible. Its energy is derived from our huge sigh of relief that the world must still be there because Gwyneth Paltrow has just dined out in it.

When Lovelock was forced to answer Richard Dawkins's criticisms on Gaia theory, he wrote a computer program called Daisyworld that showed how a planet populated only with black and white daisies thermostatically maintains itself through the respective rise and fall in each daisy population in response to dips and peaks in temperature. White daisies flourish in the warm, black daisies in the cool. Since white daisies reflect the sun's heat, then too many of them cool the planet down, allowing conditions for black daisies, and vice versa. If Lovelock's Gaia is a stable cycle of creation and destruction, a balance of rejuvenation and decline, then the two poles of mainstream media consumption, death and gossip, story and non-story, time and non-time, demonstrate a cleavage in which destruction and creation are prised apart and allowed to bloom in parallel. Alongside a morbid fascination with the end of history, we have the Always-Always Land of celebrity news.

Lovelock's neighbour William Golding told the scientist that he needed a name for his theory of a self-maintaining ecosphere, and that he should name it after the Greek goddess of the earth. There is no official name for the seemingly endless proliferation of celebrity culture's hollowed-out happenings, but one metonym that British social commentators often invoke is that of the glamour-model entrepreneur Katie Price. In the context of media theory, as an unchecked juggernaut, she represents the opposite of Gaia's equilibrium.

Lovelock was first to discover the threat of CFCs because he noticed during his country rambles that the Sussex Downs were often less visible in the distance than they had been when he would walk the same trail as a boy. From analysing the haze obscuring his view, he found that the air was silently accumulating unchecked amounts of the manufactured gas. One day recently, I realised at the supermarket checkout that Katie Price was on the covers of over half the magazines on the little rack, dominating this vista of nuts and tic tacs and newspapers. Price is, in social scientific terms, a fascinating marker of a process. Lovelock chose to analyse the Sussex haze for CFCs because these chemicals were an effective marker of uniquely industrial activity, and no one is more industrially produced than the phenomenon formerly known as Jordan. Her ubiquity has been such that it would not be surprising to find trace elements of Katie Price in the fat of Antarctic penguins.

Andrew O'Hagan has described Price as 'the nation's most frightening glamour girl', but her fearsomeness is quite different from that created by panic's storytelling. Hers is the opposite process, one of anti-narrativisation. While our modern anxieties are often based on a deranged, disproportionate imagining of cause and effect – if I step on this plane, then this will happen – Price's career soothes us with its lack of such consequence. Her cultural presence, while perpetuated by certain contemporary obsessions with mediated reality, is in fact most reminiscent of a much earlier form. Her public life is lived according to the logic of the pre-modern annal, the dominant mode by which reality was documented a millennium ago. Thus, while Gaia is an icon of classical proportion and balance, the reality in which Price stars is medieval.

In an essay on narrative forms, historiographer Hayden White gives, as an example of the annal, an excerpt from *The Annals of Saint Gall*. Here some events from the eighth century are recorded chronologically in a vertical list, each entry beginning with the year in which the event happened. Examples of entries are:

712. Flood everywhere.
714. Pippin, Mayor of the Palace, died.
720. Charles fought against the Saxons.
721. Theudo drove the Saracens out of Aquitaine.

White writes of the annal that 'All the events are extreme' and, crucially:

> Social events are apparently as incomprehensible as natural events. They seem to have the same order of importance or unimportance. They seem merely to have *occurred*, and their importance seems to be indistinguishable from the fact that they were recorded. In fact, it seems that their importance consists of nothing other than the fact that they were recorded.

The annal makes no attempt to synthesise a narrative from its list of events. With the annal it is not the scale of the event, or the factors that led up to it, but rather the fact of the event which most counts. Katie Price trades in this distinction. As with the death of poor Mayor Pippin, Price's decisions are unquestioned by her annalists; her 'splits' and her heady romances are sold off with equal vigour; her banalities are lauded simply for occurring. At more or less the same time as Sina-Inakoju's murder and Lovelock's solemn prognoses, we could compile the following excerpt from the *Jordannals*:

4 April. Price's marriage reaches crisis point?
4 April. Price 'attacks OAP over parking space'.
4 April. Katie Price 'ups security after threats'.
6 April. Psychic: 'Spirits are drawn to Price'.
7 April. Katie Price: 'I contemplated suicide'.
7 April. Price 'helps friend after horse horror'.
13 April. 'Katie Price isn't pregnant yet'.
15 April. Price given 'exploding boobs' warning.

The cruellest month indeed. As with White's example, Price's accretion of headlines displays no coherent story or dramatic arc. Unlike digital memory's capacity for assembling robust histories, the unfolding present of celebrity culture dissolves the ties between cause and effect. All that binds these events together is the intensity with which they are reported and distributed. In place of formal plotting we find a drive for list-making, fevered and unending and devoid of any logical narrative conclusion.

Price certainly senses her own position outside history. In November 2009 she returned to the Australian jungle set of the ITV reality series *I'm a Celebrity . . . Get Me Out of Here!* She quit the show early after the phoning public repeatedly selected her for bush-tucker trials, those infested challenges in which the chosen celebrities try to win food rations for the camp. Having endured several of them in a row, Price begged the public to stop voting for her, and yet the next day she was elected once more for the latest ordeal. Her confusion over why only she, out of ten other celebrity contestants, was continually nominated belies the crux of her own media phenomenon. 'Either they love me or they hate me. No idea,' she said. This ambiguity makes stark Price's function as the embodiment of pure reportage, since the engendering of love and hate, sympathy and spite, relies on the substrate of narrative: emotions usually require a story. Rather, she lies sublimely beyond the place where stories are told. As a result, her fans have long forgotten the difference between worshipping and torturing her.

Price claimed that she returned to *I'm a Celebrity* because she wanted 'closure' on her marriage to Peter Andre, whom she had met on the same series five years previously. The reason she gave for her departure (besides her monopoly on bush-tucker trials) was the fact that the jungle reminded her too much of Andre. It is

Social media offers one strategy for consolation in a present weighed down by the past's connections and stories, and hollowed out by a doomed future: to give the sense of all the moments across one's known world happening all at once. The present is shattered into a hundred incongruous pieces and laid out before us. We do of course trace stories using social media, following the Twitter timeline avidly to piece together an unfolding world event, to understand the order in which things are happening. At the level of content, narrative and significance certainly survive. But at the macro scale the form, that whole mad unravelling of non sequiturs, is freedom from the feeling of time running out. For there's always more of it; the posts keep coming. It's ironic that when the world seems at its most dangerous, when some new horror is occurring and one thinks that this could really be end-of-days, that social media provides us with a glut, a surplus of time. 'My timeline is moving so fast!' I remember one Twittering friend saying during the bloody beginning of the Syrian Civil War. In crisis the timeline grows and grows, the instant stretching and stretching to contain these shards of thoughts and fears, life moving sideways rather than forwards, where the real trouble lies.

This is the panoramic view of social media, seen from on high, where the overall form can be appreciated at its most marvellous and expansive, and from where the apocalyptic signs appear as small print. In the microscopic view, however, even the form is less liberated from time. All posts bear their age, and, taken in isolation, we can watch them set into history before our eyes. If we happen to zoom in on one post during our perusing, it can become a terrible register of time passing. Perhaps we have seen this little post arrive fresh into the world, at the top of the heap, and as we punctuate our own days with dips into this medium, we see what time has done to it. We can remember when 'This!', attached to an article about vaccinations, was eight seconds old. With the light sliding away from our afternoons, we return to it bearing the mark of four hours. 'I remember you when you were forty-one minutes!' In our heads we swear vulgarly, according to our tastes. That mini-panic for the squandered present arises: where did the day go? What's more, we know the posts are more durable than we are. For the first time in the history of a species, the bon mots of vague acquaintances are wry little memento moris.

Porn-time

H. G. Wells's *The Time Machine*, published in 1895, imagines a future extending well beyond Lovelock's apocalyptic twenty-first century, though admittedly humans as we know them are conspicuously absent. In the story, the intrepid late Victorian known as 'the Time Traveller' courses through the centuries, 'in great strides of a thousand years or more', from the comfort of his rigged-up armchair. His travels take him late into the planet's life, when earth lies dying and no longer spins on its axis, so that the same cheek now permanently faces 'the huge hull of the sun, red and motionless'. Like Shirley Valentine, he finds himself on a melancholy beach, thinking about time. The last crustacea – giant crabs – lumber along the shore and when he throttles his machine once again, they too disappear as the 'red-hot dome of the sun had come to obscure nearly a tenth part of the darkling heavens'.

Wells was a witness to the late nineteenth century's four-dimensional mania, and *The Time Machine* begins with some exposition on the debate. The Traveller, having returned home from his cosmic adventures, gathers some sceptical friends together and tells them:

> Really this is what is meant by the Fourth Dimension, though some people who talk about the Fourth Dimension do not know they mean it. It is only another way of looking at Time. *There is no difference between Time and any of the three dimensions of Space except that our consciousness moves along it.*

Once time is thought of as being akin to space, it becomes possible to imagine traversing it bi-directionally, like someone pacing up and down a room.

If the Traveller hailed from our own era, he would no doubt be branded as a purveyor of 'Time-porn'. His descriptions of the years elapsing in an accelerated blur before his eyes would never get past our hair-triggered pornographic sensors. Imagine what BuzzFeed would make of sketches such as this: 'the sky took on a wonderful deepness of blue, a splendid luminous colour like that of early twilight; the jerking sun became a streak of fire, a brilliant arch, in space; the moon a fainter fluctuating band'. Only nineteen more examples to find. The early years of this decade have seen a concerted rise of this suffix '-porn', which gives a jade-coloured vapour trail to both our innocent and pernicious amusements.

One explanation for all these porn metaphors in digital life would follow the same reasoning as why shepherd-poets would be inclined to liken all sorts of things to hills and the trees and the sky. Another explanation is that we are alert to the mediated quality of our digital experiences, a by-product of the inevitable voyeurism of going online. The metaphor is, on the face of it, spatial, referring to how digital technologies allow us to experience life from a distance. But I want to suggest that there is also a temporal reason, that all this metaphoric porn also stems from the ways in which porn's particular style of timekeeping has begun to pervade the rest of life. Each utterance of 'something-porn' betrays the tick of the four-dimensional clock, one that is not strictly within chronological time at all, but rather, like Wells's Traveller, outside that weary, single-minded trudge. Moreover, this 4D 'porn-time' appears to be one of our minor symbolic rebellions against the threat of a foreclosed future.

This racy suffix is related to the boom in non-fictional programming, as well as all the 'real life' imagery on the web. Jamie's and Nigella's cooking shows have been harnessed to the term 'food-porn', which is not merely a comment on their sex appeal. Blogs such as Smitten Kitchen, its sumptuous pictures devoid of human body parts, would also count for some as food-pornographic. 'Property-porn' was coined by journalist Rosie Millard in 2002 to describe the habit, both online and in print, of comparing your own dwelling to homes of similar value and fantasising about the wonderful places your equity might take you. There are televisual versions of property-porn, too, which trade on the vicarious pleasures of seeing people feathering or upgrading their nests. Though perhaps less mainstream, we can intuit what is meant by 'nature-porn': David Attenborough's breathless whispers would now be deemed, in the eleventh hour of his career, to coax our bestial voyeurism. More sadistically, America's appetite for sensational chat shows of the *Jerry Springer* variety have been, since the millennium, considered examples of 'poverty-porn', and documentaries such as *Benefits Street* (2014) and *Slumming It* (2010) have been accused of providing unhelpfully titillating portraits of socio-economic degradation.

In the Victorian imagination of the Time Traveller, the act of rich people going to ogle poor neighbourhoods was viewed, more literally, as touristic rather than erotic. A culture reveals much about itself by the metaphors it uses. While some of these '-porns' were conceived in the pre-digital era – poverty and food date to the late 1970s and early 1980s respectively – they have only become relatively commonplace since the late 2000s.

This recently opened porn umbrella assumes that very different types of viewing are all equally analogous to sexual excitement, and yet the metaphor soon breaks down. Millard plays with her own conceit when she confesses of one of her pin-up houses: 'We had dirty thoughts about the Rookery, which has an orchard and a moat.' A headline on the Seattle-based, environmental non-profit organisation Grist's website creates a strange hybrid of sauciness and piety: 'This time-lapse nature porn is your five-minute dose of Zen.' In 2008, a reviewer of Smitten Kitchen testified that 'The food-porn factor of this site nears illegal levels', yet no matter how glistening a wild mushroom and Stilton galette may be, it would certainly be a wet fish in the sack. The teen 'gross-out' movie *American Pie* is a better example of food-porn than a picture of stuck-pot rice with lentils. Meanwhile, 'poverty-porn' replaces the masochism of gazing at out-of-reach wonders with the sadism of class superiority, and by association casts both the other '-porns' and the real thing in unfavourable lights.

So why are all our kicks suddenly being associated with pornography? The act of browsing a series of striking images online, which may be striking for a wide variety of reasons, has been subsumed into the narrow category of flicking through a skin mag. It's fair to suggest that porn-porn, unlike the other cuckoos in its clan, has a much more conclusive endpoint: the moment, let's say, in visual terms at least, when your engrossing, busy companions of the last little while transmogrify into slightly embarrassing strangers, to be clicked away with a fickle and weary stroke. However, the four-bed Georgian town house in Margate, giving itself up for less than your London hovel, is a perpetual stud. Where then is the release in all these metaphoric sex-aids? To return to Narnia, they seem to represent a case of 'always winter, never Christmas'. It's odd, therefore, that we all know what we mean by these '-porns', despite the comparison being imprecise.

We have latched onto the '-porn' tag to describe certain, often incompatible, types of viewing, when arguably internet porn, and indeed porn in general, was the precursor of a far more omnipresent feature of online life that has to do with how we experience time online. People en masse were watching video porn sites before they were watching television shows and films on their computers. The porn movie industry was quicker than most to hitch up its wagon and pioneer cyberspace. Since then, online porn has always been as much concerned with temporal perversion as with anything else. Disease-riddled free sites, which have sent many a PC into a panting tailspin, chop up their pro bono clips into brief glimpses of time. In this way, they present the material with the same lack of emphasis on narrative coherence as on bodily coherence: the porno is packaged into a series of time fragments just as the sex itself is so often shown as a series of collisions between fragments of bodies. Cause and effect, which has to do with the linear, sequential flow of time, has never been one of porn's fortes. Behind the safe walls of pay-per-minute sites, which provide condoms for your hardware, you watch the sex but consume the seconds. In this harsh, post-crash climate, time is therefore of the essence, and so, even with a full movie available, it becomes good economics to cut to the chase.

To a certain extent, however, when it comes to porn, time has always been of the essence. Pornography buffs, going back to those slumped, old-school aficionados with remote control in hand, inevitably slice and dice the porno's timeline into a catalogue of useful segments. But now this behaviour has become a primary feature of digital life. With the advent of the laptop cinema and the ephemeral time bar at the bottom of the screen, even the primmest among us can interfere with the narratives of our Netflix programmes, movies or YouTube videos in a way that our non-horny laziness prohibited in the blind, whirring days of fast-forward and rewind. How often in viral video posts are we instructed something like: 'go to 1:28 to see how this little girl destroys the debate on world hunger'? When online viewing involves revisiting that which we've already seen, such as my impatient pawing at *The Box of Delights*, we can now home in on our money shots. Sliding the time bar's toggle back and forth like a trouser-fly, we discover we've watched an old favourite film in ten minutes of self-edited highlights.

It's therefore not so much our spatial relationship to the online imagery itself – the beddable Rookery or the toothsome *boeuf en daube* – that is metaphorically similar to porn, but the way in which we're apt to spend time online. You only have to admit to some of your own browsing habits to feel some of our shared urgency and restlessness towards all sorts of digital content, an urgency which, before technology enabled it, was more typically linked with the eagerness of the sexually aroused. If porn was once a private wing of life, very much segregated and distinct from the daily grind, the pace and rhythms of porn viewing are now pervasive. This is, I believe, why we so easily dub things porn-esque that in reality bear little resemblance to erotica. More and more we move to the porno's tempo, a boom-chicka-boom bassline throbbing subtly through cyberspace. Netflix is related to skin flicks not because it gives us access to *The Life of Birds*, nor allows us to get any other of our non-erogenous rocks off, but because of the command we have over its timeline, our ability to rove back and forth among a programme's most potent scenes. That's not to say we don't watch things in order, from start to finish, as in the old style. Of course we do. The dream of Jack Horner, Burt Reynolds's porn-director character in *Boogie Nights*, is to make movies that people would watch to the end, even after orgasm, and we haven't yet lost this stamina. Like many of you, I watched *The Killing* from start to finish, long after I'd shot my sweater-porn wad.

Developing a system of time travel is a sensible strategy for anyone asked to live in a culture that prophesies impending disasters on every scale. Simply the idea of climate change, in particular, has been disastrous to any carefree vision of the future. What a downer it is, in temperate England at least, to experience every gorgeously warm day as a symptom. Wherever you happen to be, it is easy to think that today's weather would be healthier if it were slightly different. Climate change is a bad combination of intuitive and unfathomable, and the vagueness surrounding its own timeline doesn't help with scheduling one's malaise. It's therefore fitting that, during a time when projecting oneself into the future requires seeing through a gathering smog of carbon emissions, digitisation facilitates these small, temporal mutinies. Climbing into bed with our laptops, we rove like the Traveller into the before and the after. We waggle our fingertips back and forth over the mousy rink, seeing 'trees growing and changing like puffs of vapour, now brown, now green; they grew, spread, shivered, and passed away' as we chase that barely remembered moment from *Dawson's Creek*, and the smell of 1999.

one of the funniest contributors to my timeline. Her posts are like the daily view of shop frontages as seen from the bus, whose details strike you on some days and mean nothing on others. But what happens to a person when they become scenery? One could argue that as long as we have lived in communities we have formed the backdrops to one another's lives. Those illustrated vocabulary books for children are accurate in this way, their street scenes packed with peripheral nouns: the baker, police officer, gardener, cook, cleaner, nurse, teacher, all constituting the intricate but distant patina of the outside world. The four-dimensional difference, however, is yet another warping of scale, so that the close-up has become the backdrop. An oddity of the scrolling screen is that it gives a view onto both the subjects of our laser focus and that which we would otherwise have seen out of the corners of our eyes. It's a place of full-frontal glimpses, where we encounter the periphery head-on.

This arrangement doesn't feel entirely wholesome. I suspect that my face when perusing Facebook has, more often than not, all the wonder and animation of the customs officers checking the scans of carry-on luggage at the airport. How often can those watchdogs be seen stretching their backs over their seats, between one handbag and the next, then returning without vigour to their little telly and its pastel fantasia of everyday belongings, as imagined by Andy Warhol? Does the child babbling happily in the bath deserve to be seen by my jaded, shift-worker's eyes?

Recently the stranger-friend had been having trouble with hecklers on Facebook: fights between online-only acquaintances and her real-life pals. These skirmishes took place in the threads unspooling from links to politically divisive news articles. And so she announced, sternly but reasonably, that she would be 'culling' from her friends list all those people whom she didn't know or couldn't remember meeting.

I couldn't imagine surviving this decimation. If she hadn't made her grand edict, if she had snuffed me out without fanfare, a fleet assassin, I wondered how long it would have taken me to notice. For there's a curious symmetry to her cull in that she is culling both of us. Elizabeth Bowen, in her novel *The Heat of the Day*, has a good line about the negative imprint the London Blitz left on the city and its populace: 'not knowing who the dead were, you could not know which might be the staircase somebody for the first time was not mounting this morning, or at which street the newsvendor missed a face, or which trains or buses in the homegoing rush were this evening lighter by a single passenger'. Do we notice now, when a friend slips away, our freight of followers lighter by one? There is the twinge of realisation, when a name that has glided by for years, perhaps, in streaks of dull news, has at some point stopped gliding by. How long has it been, since they called an end to the charade? You look them up, and their page is now a shut door, in the middle of which there is a grill through which no eyes peer. A notice on the door lists some formal instructions – send a Friend Request, begin all over again. It is bureaucratically convivial and meant for someone else. You are the *persona non grata* standing on a welcome mat. You might laugh to yourself and reach for your tea with its curling steam – you can't fire me; I quit. But you will remember it again, the closed door.

In the following days, whenever her Jabberwock child materialised in my news feed, I found myself returning to her profile page and checking if her friend count had been massacred. It was disheartening to see the crowded tally still intact, since it meant I'd survived nothing, and some days the number had even advanced by one or two. The extent to which I cared was strange and dismaying. One night I dreamed of this woman – moving with her through the rooms and corridors of a vast, run-down imagining of where she lives in the East End – trying, my dreaming self knew, to curry favour, at moments almost grovelling. The one small power she had over me in life migrated to the dream and became absolute. Back on her Facebook page, people were begging for mercy. 'Don't cut me off!' one pleaded; another wrote a polite, plaintive goodbye, meekly bending his neck to the blade.

Shortly after the announced cull, I met her for the second time, on work-related business. She was amazed that I'd stayed around so long, assuming that I would have culled *her* long ago, or at least muted her in some way. I told her breezily that she earned her keep; one of her updates always stuck in my mind. During the extremity of her prenatal confinement, she was craving a hearty piss-up. Over Facebook she declared her devil's bargain: if it meant she could drink, then she would pound a mouse to death with her bare fist, 'starting now'. When I told her how I thought of that post from time to time, she laughed (or rather, wrote Ha! in a Facebook message), admitting that she'd forgotten about the mouse piss-up pact. When we met in person I felt that see-saw of inequality that arises when in the presence of someone famous. I knew more about her than she of me, and yet her ignorance was a mark of prestige. Such is the imbalance between the prolific social-media-ite and the silent loiterer. Our reunion wasn't a case of 'We have so much to catch up on', but rather 'I have so much to remind you about'.

Light-hearted culling is tied to the perceived immortality of our four-dimensional selves. The slaughter is only provisional, the genocide occurring painlessly in a world of one. Otherwise language wouldn't be able to march into such bloody territories. In 2007, people were casually declaring their MySpace suicide – the termination of their accounts – which is surely a minor act in a world of limitless resurrections. Life is so abundant on social media that its metaphors can be death-soaked with apparent impunity. People are everywhere, in limitless prosperity. When social-media platforms begin to fall out of favour, replaced by more popular rivals, then their rows of inactive profiles take on a necropolitan quality, filled with hordes of the unmourned dead. Even when such networks are thriving, they are so vulnerable to progress that we feel their mortality even in the bloom of their prime. As a young correspondent for Sky News, Grace Beadle, wrote in early 2014, 'Facebook is slowly, but surely, joining Bebo and MySpace in the social network graveyard.' The animating spirits of these profiles, also known as real people, have of course moved on to haunt other sites.

Digital life is inherently suited to a language of the macabre and the monstrous. The most established example is the troll, an impressive digital metaphor in the sense that it combines the word's fishy meaning of 'to search around for' with its barbarous associations. As both verb and noun, an internet troll scours social media and website comments sections for a place to drop anchor and instigate something on the spectrum between mischief and misery. It has been argued that troll is too broad a name for this range of behaviour, which runs from pesky inanity to verbal abuse and death threats. Objections also arise to the term's folkloric flavour, suggesting that the metaphor's whimsy obscures the sadism behind it. Imagine if we collectively referred to the real world's criminals in such a fanciful way, reporting a drink-driving accident as: Toad of Toad Hall Kills Schoolgirl, 14.

The varied use of trolling, as with our panic culture of personal and global tragedies, is another example of our modern-day problems with proportion, a flattening of scale. In this case, we have the grouping of the considerable and the negligible under the same banner. Curiously, the literary troll itself invites a gathering of opposites, since these cave-dwellers, originating in the Norse tradition, are *either* giants or dwarfs. Insensible to these reproaches, 'troll' has tunnelled its way into the lexicon. As I write this paragraph, the BBC is reporting the story of 'The Troll in the President's Office', about how a notoriously abusive Twitter account allegedly originated from one of the Rwandan leader Paul Kagame's employees. This routine summoning of trolls suits journalism's taste for the sensational, but it also reflects our instinctive idea of cyberspace as containing grotesque topographies. Troll was the Vandal name for ancient Goth, and as we begin to map the digital world's emerging landscapes and inhabitants, we're increasingly turning to Gothic language to explain ourselves. It's therefore appropriate that this misanthropic creature has become an early figurehead of our cyber-Gothic imagination. But beware: it is not alone in this fantastic online bestiary. The Rwandan troll, as the BBC earnestly reports, 'may have been a ghost tweeter'.

The Gothic mode is a large, cosmopolitan canvas of the gory, the supernatural and the chilling. It deals in black magic and irrational desires, in ghoulish happenings in ruined dwellings. Scholars have in the past tried to pin the period of the English Gothic novel under glass, like a butterfly, cataloguing its lifespan as beginning with Horace Walpole's *The Castle of Otranto* (1764) and ending with Thomas De Quincey's *Confessions of an English Opium Eater* (1821). Except that the Gothic is not a butterfly, but a bat whose wings beat as pages turn, flitting over the moors of *Wuthering Heights* (1847), into the alleys of nineteenth-century urban serials such as G. W. M. Reynolds's *The Mysteries of London* (1844–8), up past *Dracula* (1897) and it is still on the wing. As an international genre of the imagination, one reason the Gothic endures so widely is because it preys on our instinctive need always to be able to distinguish between presence and absence. The animal in us likes the lines to be clearly drawn between who is near and who is far, who can see us and who cannot. Even the most civilised person feels hackles rise when someone says they are leaving and then mooches around for a further half-hour. The loiterer is breaking the rules by hovering on the threshold between home and away. The Gothic seeks to amplify our vague agitation into mortal terror, which explains why the phantom is such a Gothic regular. Phantoms are loiterers par excellence, neither present nor absent.

Digital life is primed for Gothic interpretation precisely because of its tendency to blur these lines. We have already seen the simultaneous hereness and thereness of the four-dimensional body, as well as how digitisation inverts our ideas of the solid and the insubstantial. Recall Seamus Heaney's poetic formulation of 'there-you-are-and-where-are-you?', which is a question both for our four-dimensional web-presence and for the spectre. In broad terms, the uncanny event, another Gothic staple, can be reached by two different lonely roads, both of which snake through cyberspace. The first, as noted in Airbnb's weird form of homeliness, is the road that leads us to experiencing the familiar as strange or the strange as familiar, the horror of the unhomely home. Travelling on the second road we begin to wonder what it means for something to be alive, to be animated. The mannequin, the doll and the statue are all potentially uncanny in their lifelikeness. Although inanimate, they can easily seem to possess an obscure conscious-ness. Their uncanny cousin is the corpse, which was once animated and is, often unbelievably, no longer. All of these things – the ghost, the dummy, the corpse – inhabit in various ways a threshold world of incomplete presence. As digital technologies absorb more and more aspects of life into such a threshold space, it's understand-able that Gothic imagery becomes more pervasive. The four-dimensional human, in other words, lives in a haunted house.

Stalkers and Spies

If there are trolls under the bridges, then there are ghosts in the 4D attic. In the summer of 2013, following an article in *Slate* magazine by Seth Stephenson, there was a groundswell of interest in the idea of 'ghosting'. The term pre-dates Stephenson's piece by at least three years, appearing in the *Urban Dictionary* in 2010 and referring to the practice of leaving a party without saying goodbye. As Stephenson points out, this phenomenon was named as early as the eighteenth century, possessing standard xenophobic aliases such as the Dutch or the Irish farewell. With public taste since having veered away from these casual digs at other nations, we have had to concoct a new phrase for the unheralded departure. Ghosting is certainly consistent with the spectral ambience of the times. We often ghost during digital interactions, as seen in the text exchange with no formal farewell, or when someone drops silently away from a social-media thread. But in addition to these mysterious comings and goings, there's a new and ghostly aspect to social relations that has to do with our incomplete transition into online camaraderie.

All this socialising has created four broad categories of friends: those planted solely in the real world and who never follow your online exploits; those with whom you interact in the real world *and* online in glorious balance; those strangely intimate friends whom you only know through online media; and, the spookiest group of all: those whom you know in the real world and who are aware of your online exploits, as you may be of theirs, but with whom you rarely if ever interact online. This last category arises from the rift that has occurred within that large population who dutifully signed up for the social-media experiment and which then became segregated into those who communicate with others and those who choose to lurk watchfully.

Social media isn't just one thing, and some platforms invite lurking more than others. Facebook is especially haunted by watchers, their interest in others so often propelled by tedium, which no doubt results in the ill-spirited conclusions drawn about those whom they like perfectly well in the flesh.

If you are one of the watchers and are friends with a prolific Facebooker, for example, you may have noticed something retro about your catch-up drinks. Your prolific friend tells you their news, and even their latest interests and opinions, but so much of it is half familiar, reminiscent of what you already know. You suspect that there has been a leak, that some mole has already informed you about the carnage and debris of the big night out, the song that has been stuck in their head or the picturesque skies over Devon. You look across the table and realise that the mole is looking back at you: your friend has leaked themselves!

As a result, these friends, sitting with you in what we used to call real time, resound like echoes. Although they're with you in the present, they can seem further back in time than the cutting-edge version of themselves that live-streams into your pockets. This lag between virtual image and real-world self is comparable to the dissonance in Oscar Wilde's late-Victorian Gothic novel, *The Picture of Dorian Gray*. When Dorian casts the actress Sibyl Vane aside because her desire for him has spoiled her performances (with which he fell in love), he returns home to find a change in his portrait: 'The expression looked different. There was a touch of cruelty in the mouth.' The next day he learns that Sibyl has killed herself, and realises that the portrait knew before he did, that it 'was conscious of the events of life as they occurred'. One of Instagram's official aims is 'to allow you to experience moments in your friends' lives through pictures as they happen'. We all know that it's not uncommon for Instagram to consume our dinner before we do, or for our smartphone to see the concert first. Our digital portraits crest the wave of the moment, while our sociable selves, the ones still tuned to a somewhat outmoded notion of 'catching up', can fall behind the times. The real, biologically up-to-the-minute 'me' thus becomes a ghost of my online self, clanking the chains of some bygone rant that played itself out on Facebook days before.

When Dorian shows Basil Hallward, the portrait's painter, the supernatural changes to the creation, Dorian's expression is itself an after-image of the first alteration. 'Dorian Gray smiled,' Wilde writes. 'There was a curl of contempt in his lips. "Come upstairs, Basil," he said, quietly. "I keep a diary of my life from day to day, and it never leaves the room in which it is written. I shall show it to you if you come with me."' Dorian's initial horror of owning a portrait that registers his sins is analogous to our own anxieties over the purity and virtue of our online profiles. People applying to universities or for jobs are increasingly liable to think of their web-presences as potentially incriminating. A social-media profile can become a book of evidence against one's moral rectitude. The young are routinely warned that the behaviour recorded by the digital gaze could jeopardise their dreams. Prospective employers are checking up on their online lives, they are told. Unsurprisingly, paranoia revels in this environment. We become like Dorian, worrying that his servant, Victor, will discover the shameful portrait: 'Perhaps some night he might find him creeping upstairs and trying to force the door of the room. It was a horrible thing to have a spy in one's house.'

In response to this panoptic surveillance, apps such as Ajax are emerging to 'clean the hell out of your digital life', as well as a host of agencies providing 'audits' of our social-media pages. These services scan our digital pasts and remove unsavoury jokes or associations, deleting crass links and drunken leers. Pictures featuring the white jut of a cigarette are quietly removed. Unlike in Wilde's tale, we're ultimately both the painter and the subject, and so we can daub away compromising or otherwise unflattering expressions of ourselves with a simple un-tagging. Yet, whether or not we choose to preserve only our most beautiful features in this online portrait, built into the practice that prioritises documenting over experiencing is the pause, perhaps imperceptible, which makes life happen to the portrait first. In this sense, it is older than we are.

Our intuition for the Gothic qualities of digital life is also reflected in the explosion of the word 'stalking'. When the young talk to each other about social media, they routinely use a vocabulary that moves between espionage and obsession. 'I totally Facebook-creeped you last night,' they will confess cheerily, meaning that they browsed, in a manner that felt excessive, through their friend's profile. There's an assumption that no one is very many quiet steps away from being a stalker. But all of these spy and creep allusions amount to a kind of bashfulness, as if we're not ready to admit to the voluntariness of our friends' revelations. We have the sense that, in order to be seeing what we're seeing, we must have somehow broken in, a feeling that mistakes a display case for a vault. This confusion is one outcome of reversing the peephole: the intimate view into the private lives of others can seem ill-gotten, even illegal, despite the fact that our vision has been sanctioned, indeed invited. We can't believe that such intimacies, once held inside the vault of privacy and decorum, could possibly be on display. Our three-dimensional manners have yet to catch up with four-dimensional norms of revelation.

That these confessions happen at all suggests a certain light-hearted guilt that needs to be voiced, since in most instances such creepery is the perfect crime. No one but yourself need know how long you spend clicking around someone else's online particulars, sifting through photographs and eavesdropping on long-cold conversations. A 4D bogeyman arises with every mention of the apocryphal app that lets Facebook users 'see' who has been visiting them. What could be more horrific than such a dusting for finger-prints? While this ongoing threat may discipline us to some degree, there's also an old-fashioned decorousness at work. We are compelled into politeness-by-proxy, which means we shouldn't outstay our welcome, or rummage too deeply into the bedside drawer of our friends' timelines, or return too often, or read too far into the diary left open, by-accident-on-purpose, on the table. The old mores of civil society persist. It is after all chivalrous to think of oneself as a stalker rather than one's friend as an exhibitionist. Is it creepy to go to a well-advertised show at a gallery? Do I stalk Matisse around Tate Modern? By Gothicising our own and each other's motives, we obscure the potentially more morti-fying thought that we have opened up our lives to be seen.

The Haunting of Wiltshire Farm

The digital age's concern, in multiple senses, with how we are being viewed online reinforces its Gothic inflection. Have you ever been in an unfamiliar room, perhaps left for a moment by your host, and as your gaze skims across the mantelpiece and up between the curtains, your mind trying to understand what it might be like to inhabit the mind that made this room, you notice that all the while a cat has been watching you from among the cushions? Even if this silken ambush has never happened to you, it's not hard to imagine the little jolt of unease and betrayal at having been looked upon during a supposedly private moment. As a species, an aspect of our evolutionary fitness seems to include a useful dose of paranoia. The predisposition to feel that we're being stared at – our napes chron-ically on the verge of tingling – would in theory give us an advan-tageous readiness for danger. Paranoia, unsurprisingly, is a temperamental servant, one that is apt to perceive eyes where there are none. The Gothic exploits our narcissistic and jittery inclination to feel the subject of attention. Even Dorian Gray, confronted with the breathtaking impossibility of an aging portrait, remembers to notice that 'it was watching him, with its beautiful marred face and its cruel smile . . . Its blue eyes met his own.' Gothic castles are notoriously furnished with portraits whose gazes follow the nervous visitor, but the question of who or what is looking at us has become central to the unfolding ethics of digital life.

The first scene of Wes Craven's *Scream* franchise is a modern illustration of the typically Gothic power imbalance based on sight. These films are well known for their self-conscious use of the horror genre's clichés. In this opening, Drew Barrymore plays the blonde maiden who loves movies such as *Halloween*, featuring 'the guy in the white mask who walks around and stalks babysitters'. In postmodern fashion, she is discussing her favourite scary movies over the phone with one of her imminent, white-masked killers, who instigated the chat by pretending to be a wrong number. You might think that all this irony would smother any chance at a fright, but this coy sequence manages also to be terrifying because it earnestly uses a basic Gothic technique of uneven vision.

Barrymore's character, Casey Becker, is in a house 'in the middle of nowhere' and she paces around its illuminated ground floor, crossing abundant windows that reveal nothing but the blackness of the night. She is therefore in the vulnerable position of being on display to a potential legion of invisible observers. Besides our awareness of what sort of film we're watching, the camera primes our uneasiness by being itself an invisible eye that follows Becker from room to room while she banters over the phone with her killer. But the first pulse of pure terror hits when she asks, 'Why do you wanna know my name?' and the coldly wooing male voice replies, 'Because I wanna know who I'm looking at.' The knowledge that she is being watched by someone she cannot see sends her locking every latch she can find. She peers out through various windows, squinting into the darkness, while the stranger on the phone continues to taunt her with his optic advantage. 'Can you see me?' he purrs, and when she asks what he wants he admits that his aim is to further his visual domination over her: 'I want to see what your insides look like.' The menace of this scene comes not simply from the caller's 'trolling', but from his and Becker's respective positions within the darkness and the light.

In Poe's story 'The Pit and the Pendulum', a man condemned to a shadowy cell by the Spanish Inquisition feels that 'My every motion was undoubtedly watched.' Aside from the human gaze, there are the non-human watchers to contend with. Poe addresses this perspective, too, in 'The Fall of the House of Usher', where Rodney Usher feels the decaying house exerting its influence on him: 'an effect which the *physique* of the gray walls and turrets, and of the dim tarn into which they all looked down, had, at length, brought about upon the *morale* of his existence'. The Gothic mode relishes the unsettling politics of vision, and explores how power can be wielded and terror incited by having someone feel scrutinised and influenced by forces that they themselves cannot see. One of Roderick Usher's creepy paintings illustrates a vault that seems to be windowless and underground, and yet is illuminated by an unknown source: 'No outlet was observed in any portion of its vast extent, and no torch, or other artificial source of light was discernible; yet a flood of intense rays rolled throughout, and bathed the whole in a ghastly and inappropriate splendour.' Poe's narrator tells us that 'if ever a mortal painted an idea, that mortal was Roderick Usher', and in the image of the floodlit vault he captures the essential Gothic idea of hideous illumination. This painting is a schematic for the horrors of being helplessly lit up and available to scrutiny from unknown eyes, a schematic that in our times has been exploited by various visionaries. You know who they are.

'17 others viewing this hotel right now', the travel site Expedia coolly informs us. It's understandable that we have spies and stalkers on our minds, since, as we live through our digitised hours, we know for certain that we are being tracked, though not generally by our friends or our employers. The commercialisation of the internet is no small part of its eeriness, daily ushering us into a world of surveillance, both blatant and subtle. Online merchants are shameless stalkers, and are also keen to reveal how many other people are eyeing their prizes. eBay tells us that the 'Small Vintage Antique Shabby Chic Style Chest Cupboard' has '132 watching'. Scores of unseen eyes, peering out of the forest, surround these online commodities and indeed become another dimension of their worth. For obvious reasons of competition we're made to feel the presence of our rivals, and somehow they're not just looking at Shabby Chic; they're looking at us, sitting on our couches, the windows black with the night, looking at Shabby Chic. The horror!

One of the most famous advertisements in literature is a pair of eyes, belonging to the decrepit hoarding in *The Great Gatsby*. Standing guard over the Valley of Ashes is the optician Dr T. J. Eckleburg, who has been weathered by generations of teachers into a smooth pebble of significance: the eyes looking down over twentieth-century civilisation no longer belong to God but to a salesman, whose interest in us is economic, not spiritual. Fitzgerald's image has lasted because of the note of finality that it sounded. What else is there to say, once God is dead and our new idols come from the barren world of commerce? A ruin, strangely, is more permanent than a palace. We know from Percy Shelley that most of Ozymandias' statue has disappeared, but somehow we feel that its foot will survive for ever as a spokesperson for vainglory, on those 'lone and level sands'.

Even ruins, though, aren't for ever. At some moment in this new millennium Dr Eckleburg tipped over, face first, onto the earth.

Glasses to ashes. A new, digital billboard replaces him, and the student of *The Great Gatsby 2.0* will learn that we're no longer watched over by the eyes of God, nor the eyes of a strange old optometrist, but by the eyes of those we know best. Eckleburg's spectacles have been upgraded to the double Os of Google, out of which peep, on rotation, our loved ones' eyes. Among Google's evolving terms and conditions, as of October 2013, was the following promise: 'To ensure that your recommendations reach the people you care about, Google sometimes displays your reviews, recommendations and other relevant activity throughout its products and services.' We've climbed *inside* these adverts as part of Google's 'Shared Endorsement' scheme. If you review something using your Google account, then your avatar will appear beside it when one of your friends in Google's social network, G+1, brings it up on a search. This is where we do haunt our friends: appearing in the midst of their hunt for good Italian restaurants in Chicago.

You can of course opt out of this amateur endorsing, though the process is not a simple, accessible off-switch, and Google does try to dissuade you till the last. 'Are you sure?' it bargains from the brink. 'When you disable this setting, your friends will be less likely to benefit from your recommendations.' The tone of urgency and diligence is curious here, as though Google is relaying vital intelligence by Sherpa over a mountain pass. For although we're encouraged to feel increasingly connected, we're also made to feel that we're each a thousand miles apart, living in a world where word of mouth has been lost on the wind. Those who profit from our connectivity have the double task of continually emphasising that such connectivity is both natural and fragile, that it can only be maintained through their mechanisms of perpetuation, and in that way they justify and normalise this haunted-house living.

Besides these explicit reminders of the digital gaze of others, the general style of online advertising does nothing to assuage our inkling that we're being stalked. One evening in bleak midwinter, I was the victim of a haunting. I was lying on my bed feeling lonesome and lazy and bad, cheering myself up with a YouTube clip of Sarah Silverman that I had already seen. I was beginning to feel consoled and in good company, when something in my peripheral vision made me uneasy before I knew why. To the right of the little minimised YouTube player was an inert poster advertising Wiltshire Farm Foods. I hadn't thought about Wiltshire Farm in some time, and indeed I had tried to forget the place. A year previously I must have googled this company to order some individual Bakewell puddings and apple custards, their soft meals being one of the few foods that could outwit my dying father's oesophageal carcinoma. The two sloping hills of their logo, though I was once grateful for them to arrive on the side of a delivery van, summoned up a terrible time of worry and pre-emptive grieving. I could smell the sweet steam that would rise up from the fork-stabbed plastic. I could hear my father scraping at his bowl, the look of ease and trust on his face and the pleasure the sweetness gave him as he pursed his lips to swallow. He would purse his lips like that last thing at night, when we would feed him a trembling spoon of laxative and settle him down. 'That's nice,' he said to me once, the first time he tasted his new peach-flavoured prescription.

So there this ad was, hovering to the side of Sarah Silverman, who was now frozen in my disbelief. For anyone suffering any sort of bereavement, the world is full of landmines and this is not necessarily the world's fault. Moreover, if I want Sarah Silverman to be digitally available to warm my midwinter nights, then I should expect to suffer some promotional material. 'If you want unsponsored cheer,' the Monopoly man would say to me, his eyes twinkling, 'light a candle.' Fully fledged young 4D adults find it natural and efficient for ads to be (in the cosy language of marketing) 'targeted at you'. How crude it must seem to them now, this idea that television commercials used to come at us, in comparison, like grapeshot, rather than lined up through a sniper's sight. I myself would rather not have a red laser dot flickering across my forehead whenever I go online.

Leaving the thin skin of the grieving aside, there is something more generally unsettling about these knowing adverts. The idea of commerce becoming sentient is the stuff of schlock horror, and was comically tackled in *Ghostbusters* with the giant Stay Puft Marshmallow Man stomping through Manhattan. At least this lumbering, grinning advert displayed a sort of indiscriminate violence; you never had the sense that he was psychically connected to individuals in the crowd, or that he could manipulate their secret wishes. Google's attempt to predict one's desires is in fact a cack-handed version of a more terrifying, Gothic motif. In the Gothic novel, places are alive with a malicious psychic energy, and the heroine is not infrequently petrified by the materialisation of signs and portents that seem to speak directly to her innermost thoughts. This primal sense of being watched, indeed stalked, by a super-human intelligence is one of the mainstays of the genre's terror.

It continues to unnerve me that the walls of cyberspace are decorated with watchful portraits from my past, hung with framed, ancestral searches for unaffordable homes and Superga plimsolls. Gothic writing transmutes the fear of being spied on into uncanny coincidences and the taunting insinuations of some unseen and malevolent power. These fiendish demons prey on their victims' past shames and vulnerabilities. The Gothic is a form in which watchfulness declares itself through old sins emerging from the obscurity of the past and into the glaring present. Being tailed by an advertisement comically replicates the Gothic effect that some overlord is keeping tabs on you, and moreover mimics the workings of conscience. *I Know What You Shopped For Last Summer*. Google ads sit guiltily even when you and they are innocent. More steely people may argue that once you know that this commercial pursuit is simply how the internet funds itself, then there should be no eeriness whatsoever. Indeed, if you're in a less macabre mood, these often misjudged attempts to deliver bespoke adverts have the pathos of a schoolchild memorising and parroting the enthusiasms of a more popular class-mate. And yet we can't escape the fact that this model of advertising naturalises a Gothic mode in which we are exposed, illuminated by ghastly and inappropriate repetitions of our idle time-killing.

'Download some ad-blocking software,' a friend tells me, 'it makes the internet look like the BBC website.' But our relationship to such software is its own kind of chase scene, a cat-and-mouse game whereby advertisers continually look for new ways to circumvent the blocks. That I should have to download a program in order not to be pursued in this way also suggests the unsettling primacy of commercial interests in our daily lives. The billboard shouldn't be the default; a poster should not pre-exist a blank wall. In my three-dimensional London life, as another double-decker bus loses its classic blush to an Adidas ad, there are several self-evident justifications for this position, both moral and aesthetic. On the four-dimensional front, in the war to save unbranded space, at added stake is this commercial Gothicisation of our online experiences, the abstract feeling that we can't outrun our pasts, that as we navigate through cyberspace we'll fail to escape the telltale Superga.

Through the Keyhole

In my last year of university, my roommate had a jock girlfriend. Now and then she would drop by our student house after an evening volleyball practice and sit at the kitchen table with wet hair. I've tried to remember something specific she said to me, but every scene is corrupted and all my memory hears now is the tone of her voice stripped of words. She smiled and laughed easily, and had a perfect lightness of touch for an occasional visitor. I think at least once I bumped into her on the campus lanes, perhaps walking with a teammate, the pair of them, perhaps, with their hoods up. Or perhaps she sat at our kitchen table with her hood up. I see her now, hooded and broad-shouldered, one arm around my roommate. I sometimes confuse her with an earlier girlfriend of his, partly because he liked girls who shared his colour scheme. For this reason she is prone to appear, impossibly, in my memory of one of our Spring Formal photographs from the wrong year, like a foil in an identity parade. A decade later, on the last night of my Canadian Christmas holiday, my ex-roommate drove through the snow to meet me at a bar. 'She's going to die soon,' he said of his old volleyball girlfriend, during a pause. 'She sent round an email.'

As it turns out, I knew her better when I didn't know her. After returning home I started to watch her, via her Twitter account. Although I was only seeing that which she had willingly made public, nonetheless my phantom legs ached from crouching at a keyhole. I knew that this wasn't the healthiest of pastimes, but internet guilt is often anaemic, and I didn't have to defend myself to anyone. She was now an academic, but had kept up her jock ways. During her palliative therapy she defiantly kept training, and raised thousands of dollars for cancer research in the process. Throughout that year I semi-regularly checked in on her. She began to walk the periphery of a habit, a sideshow of my boredom. Her will to strengthen herself smashed against the force steadily weakening her. There were photos of her exercising her poor cachectic body on a treadmill, and she increasingly bore the family resemblance shared by chemotherapy patients. Her approach was that of the athlete fighting a rival, but a torn muscle, to which the treatment had made her vulnerable, halted her regime. Gradually, her tweets about her condition charted the growing muscularity of the disease. She recorded her increasing capacity for sleep and described how walking up a flight of stairs or taking a shower exhausted her. It was an inverted diary of fitness training.

In many western societies it has been customary for the dying to be cocooned inside their last days. Ranks close around them, and more remote acquaintances await the news from a distance. For these marginal figures, a fear of passing too closely into death's orbit meets a fascination with the grim process at work behind the door of that shut-off room. James Joyce dramatises this petrifying ambivalence in his short story 'The Sisters', in which a boy anticipates the death of a priest with whom he was friendly:

> Night after night I had passed the house (it was vacation time) and studied the lighted square of window: and night after night I had found it lighted in the same way, faintly and evenly. If he was dead, I thought, I would see the reflection of candles on the darkened blind for I knew that two candles must be set at the head of a corpse.

The priest has had a third stroke, and now lies behind a window screen that projects nothing of his declining body's terrible industry. This steady, inscrutable glow distances the outsider, and the young narrator tells us how

> Every night as I gazed up at the window I said softly to myself the word paralysis. It had always sounded strangely in my ears ... But now it sounded to me like the name of some maleficent and sinful being. It filled me with fear, and yet I longed to be nearer to it and to look upon its deadly work.

Today we can indeed be nearer to this deadly work. Social media and blogs have allowed us behind the screen, and have given us access to the emotions and observations of the dying, which offers a lovely alternative to the traditional options of sequestration or a brave face. We are granted a form of communion that neither exhausts nor embarrasses either party. If they wish and are able, the dying can tell us what dying feels like, or they can enjoy the freedom of interactions that are not somehow inscribed with the signs of their disease. Here the disembodiment of the internet is an irreproachable asset, allowing people with depleted physical energies to have a robust presence alongside the most sprightly of users, and to find communities of others who share in their troubles. Indeed mortality's neurotic separation from daily life has been seen in the privileged West to be part of the malaise of the late twentieth century. The film *Fight Club*, based on Chuck Palahniuk's 1996 novel, explores this repression in Jack and Marla's obsessions with various support groups. 'When people think you're dying, they really listen,' Jack explains. As tourists among the gravely ill, they carve up the weeks' sessions between them:

Jack: You can have lymphoma, tuberculosis and—
Marla: You take tuberculosis. My smoking doesn't go over at all.

They agree to split custody on a shared favourite: ascending bowel cancer. 'You get it the first and third Sunday of the month,' Jack suggests. 'Deal,' says Marla. Since Jack and Marla's day, the blogosphere and other forums have offered a version of this consolation by unveiling and making public the very personal trajectory into non-being.

The volleyball girlfriend died while I was in Canada, once again for my winter break, in a hospital that was a short drive from where I was staying. I had been worried about her because she hadn't tweeted in two weeks. I began counting days between her posts, to see if such gaps were normal. For me, her tweets were a pulse, and the fact that they had stopped was like watching her flatline. She had, after all, gone behind the screen. My old roommate was away with his wife and children, and in any case no one knew that I had been keeping this intermittent, year-long vigil. Eventually a search brought up an article in a local newspaper announcing her death. I kept returning to read her last tweet, which ended in a stream of vivacious exclamation marks. It was unexpectedly strange to see her in full linguistic health, conjugating verbs and spelling correctly (i before e), and to think that within a few days she would slip away from language for ever. Once again like Wells's Traveller, I sat in my viewing seat and steered myself up and down the timeline of her last months, purposely driving myself into anguish, watching her pitches of hope and fear. I became an unofficial mourner, an unasked-for mourner, an unknown mourner. Mine clearly wasn't a crippling bereavement, but it was certainly disproportionate to our relationship. Grief might generally be described as the return journey of love, and as such has a symmetry of extension and intensity. Since I hardly knew this dying woman, I'm not sure where my mourning came from. My feelings didn't match my proximity to her in life, and so bear the opposite of the vampire's affliction: a reflection without a physical source.

The only possible source is my immaterial, voyeur's relationship with her online, which is increasingly becoming a way in which we engage with the dying. The public intimacy of the blog form is suited to a journal of ill health and decline, and both writer and reader have much to gain from this therapy. For the dying online diarist, people like me form part of a ghost audience, sensed but unseen and presumably, hopefully, of some comfort. Those who record the last weeks or months online are sewing an eventual diaspora of grief that is uniquely indigenous to digital life. We worry now about how our data is being dispersed, how parts of our lives are being raided without our knowing, only to manifest in multiple elsewheres. In the case of bloggers who make their own forthcoming deaths at least partly their subject, they will potentially be mourned for by an unknowable population of strangers around the world, who may be following their entries in real time, or discover them years after the death itself. The ethics of this cyber-bereavement, however, are complex. Does my faint but enduring sadness for my roommate's girlfriend, who was practically a stranger to me, add to or detract from her legacy? Am I merely feeling bereft of a pastime, deprived of this particular habit of digital distraction? I certainly mourn for her in a way that I would not have, had I simply heard the news one day, on the grapevine, rather than following her, as far as I could, down that road.

Professional mourners have for centuries been like lugubrious laugh-tracks; their pre-ordered wails and moans encourage others to act likewise, as well as testifying to the deceased's character and, in some cultures, expediting the transition to the next world. Subdued variations of these hired lamenters were the 'mutes', who were paid to stand still and silent outside the house of mourning, publicising the death with their sad countenances. In mid-Victorian London, mutes were controversial figures, mocked by Dickens in *Oliver Twist* and described by the writer William O'Daniel as being 'composed of the very lowest of the low; they are generally as drinking, gambling, and murderous a set of men, or devils, as ever abode within the walls of a jail'. The mute mourner takes on a new shape in our four-dimensional shift. If the business of dying is played out in cyberspace, offering an unshaded window into the process, then all such public deaths will be attended by a flock of this new kind of remote, digital mute, the Weeping Tom. They are in fact the quintessence of mute mourning, since their grief is both unspoken and unknowable. They are silent and invisible. Like their predecessors, they are emotionally removed from the deceased; it is typical that they never meet their subjects. But are they the lowest of the low?

It is certainly true that face-to-face engagement with a terminally ill person is demanding and rewarding in ways that cloaked online vigils cleanly evade. The latter are perhaps closer to the one-way bereavement we feel for fictional characters, and if so the question becomes whether aestheticised grief is a step away from or a step towards real experience. Social media can encourage us to think of people as a sort of casual entertainment, as something which we can dip in and out of, and sometimes flick to mute. The annal form doesn't demand loyalty. Gossip has always treated people as casual entertainment, but with gossip the pleasure derives at least as much from the over-the-fence exchange as from the content itself. Online a standard amusement is the solo gossip session, where we gossip to ourselves about news that, in any case, comes straight from the source. But the online record of dying is one case in which we shift from the casual annals back to a narrative that lives, touchingly, in time. This record can't help but be literature, irrevocably infused as it is with a sense of an ending. And so the ethics of cyber-mourning intersect with a defence of literature as broadening our moral capacities. Put simply, literature coaxes us, by a trick of perspective, to empathise with people who are resolutely not ourselves. It transports us into the minds of others. In the case of the dying tweeter, we know that there is a human behind their inevitable artwork, but it is the artwork that brings the silent viewer closest to this perishing humanity. The Weeping Tom is thus offered a space in which to experience genuine grief, whatever that might be.

It is easy to be suspicious of four-dimensional mourning. After my roommate's girlfriend died, someone who had been sending encouragement during the final weeks tweeted the death. By the time I monkey-barred onto this certified mourner's Twitter profile, the RIP message was third from the top and cheek by jowl with a newer, slice-of-life observation about jelly beans. Is it callous or liberating for death to be thus enfolded into the mundane? Although Twitter dates each tweet, the overall effect of a timeline leaves little room for solemn pauses. Depending on how the tweets fall, they can produce a bipolar strobe light of emojis. In this sense, digitisation can challenge our old instincts about the natural lifespans of emotions. In the past, a gesture of emotion that is either inappropriately brief or abnormally long-lived has been cause for alarm. A fixed facial expression, for instance, is a shortcut to horror. The sinister quality of clowns is well known, and derives largely from their ceaseless, crimson smiles. Someone who projects a singular emotion regardless of circumstance is naturally unsettling, since this emotive monotony suggests a void of empathy.

You would hope to knock the grin off someone's face, particularly if you're cowering at their long, rubbery feet. For this reason, too, dolls and statues and any masked creature are fright-night regulars, uncanny in their unflinching leers or mournful pouts. But equally deranged and deranging is an unnatural pitching from one expression to the next. In soap operas, for example, it is a villain's duty to morph a sympathetic frown into an over-the-shoulder or sideways smirk of ill intent. Likewise, in the world of evil, repentant sobs glide into the emotive opposite of maniacal laughter.

Social media can produce a version of this unsettling flux. Stark lurches in content from one tweet or post to the next inject an unintended tactlessness into a timeline. In 1901, the social scientist Georg Simmel wrote an essay called 'The Metropolis and Mental Life', in which he describes a particularly urban state that he refers to as 'blasé'. He argues that this condition is caused by an excess of sensory stimulation that a city street inevitably hurls at someone walking down it. What occurs is an 'intensification of emotional life due to the swift and continuous shift of external and internal stimuli'. Blasé arises not from caring too little, but from having to care too much about too many different things, and is for Simmel 'the consequence of those rapidly shifting stimulations of the nerves which are thrown together in all their contrasts'. Because the coding of social-media programs isn't nimble enough to convey varying degrees of gravity, it delivers a vast array of information to us in a uniform manner. If this information is read in a certain way, then, a flattening occurs whereby death and jelly beans occupy the same amount of space. Although most people would protest that they are able to distinguish between the momentous and the trivial of social media's content, it is nevertheless true that the medium itself encourages a homogenising of these extremes.

If you want to get to know someone's true character, it would be a wise idea to observe them incognito as they say goodbye to a third party. People often leave each other with smiles on their faces – something to do with the mild hysteria of parting – and if I'm in a park or other public place I sometimes like to watch how long a parting smile lasts on a stranger's face as they walk away from their companion. I feel soothed by a slow dissolution, a cloud turning from one shape to another. If the goodbye smile drops like a stone when the back is turned, then I know to steer clear of that person. But in the online world we are often robbed of these gradations of sentiment. The edges are sharp, the gear-shifts jarring. Granted, on Twitter certain topics go on runs within people's timelines, enduring more than others, and these conversations do convey a fluid arc of feeling rather than a mosaic of discrete and unconnected reactions. And yet because these threads appear like collapsed accordions, the medium does promote a fragmented perspective that, to the old-fashioned, can infect the whole with an unappealing glibness. It's possible that future generations will not feel such pangs at these juxtapositions, since they will be native listeners to the jazz of social media, and understand that the sincerity of a mournful note is not undone by the lively riffs that precede and follow it. And perhaps it is death, rather than us, which has always been and will always be tactless, intruding as it does in the midst of life. When the little boy in Joyce's story reads the priest's death notice, pinned to his door on a crape bouquet, he looks around the street and is surprised that 'neither [he] nor the day seemed in a mourning mood'.

Zombie Smiles

While professional mourning is culturally alive in the Middle East and Asia, a globalised economy is emerging in which smiles, rather than tears, are for sale. The mechanisms of online capitalism have created a demand for the professional admirer, who has turned the paid mute's frown upside down. And yet the uncanny arises here too, for the smiles are as inanimate as those in a portrait, and the economy in which they are exchanged is an economy of dead bodies. In a world where we have to formalise our friendship with our local dry cleaner's in order for this establishment to seem credible, value has sprinkled its golden dust over a business's quantifiable social-media popularity. Wherever value goes, of course, the counterfeiters are never far behind, and so we find the rise of companies such as Intertwitter offering a standard range of simulated endorsements: Followers for Twitter and Instagram, Facebook and Instagram Likes, Retweets and Tweet Favorites. We've moved from the religious injunction against false idols to the modern phenomenon of false idolisers, who represent another aspect of digitisation's Gothic uncanniness. While the voyeuristic empathy of the Weeping Toms complicates our notions of humaneness, these False Friends are as inhuman as zombies.

In December 2014, the *International Business Times* reported: 'Kim Kardashian Overtakes Justin Bieber on Instagram Following Fake Follower Purge'. How few of those nouns would have meant anything to anyone, ten years previously. The year before, a report claimed that fifty per cent of Bieber's Twitter followers were fakes. Following Instagram's wintry purge, celebrity news sites announced how the rapper Ma$e, a pre-digital 1990s star, deleted his account after losing 1.5 of his 1.6 million followers to the zombie clear-out.

Like the most effective salespeople, Intertwitter advertises both a threat and a solution. The menacing premise is that if you want your social-media profile to enable your business or reputation – an artist selling work on Instagram is the charming example they give – then the work itself is irrelevant if its follower count is too low. The meritocratic possibilities of social-media exposure are understandably not part of the pitch. As with Google's endorsement fever, it is as though word of mouth has been gagged and all those friendly viruses have called in sick. With the icy pep typically reserved for characters in sci-fi dystopias, Intertwitter tells the poor artist to fear not: 'We will give you the likes and followers you deserve and help you grow your social presence!' These words should surely be spoken over a tannoy in an abandoned space-mall, just as a police phone box laboriously shimmers into view.

Although the idea of manufacturing computer-generated supporters is both our cynical present and the stuff of dated visions of the future, as an economy of dead bodies it also has, like the mute, a nineteenth-century flavour. In Britain in this period, a corpse was a commodity because of the medical knowledge it held. Resurrectionists were the nocturnal traders in dead bodies, which they removed from graveyards and sold to scholarly surgeons. Any ivory-tusked creature, or indeed any prisoner in a privatised penitentiary, will attest to the dangers of being on the wrong side of economic value. Entrepreneurial ruthlessness reached a vivid peak in Edinburgh in 1827, when William Burke and William Hare began killing people in order to sell them as cadavers. The merger of social media and commerce has found a new way to give value to the inanimate, but it has required some cyber-resurrectionism. Like Mary Shelley's Gothic creation, whose inventor 'collected bones from charnel-houses and disturbed, with profane fingers, the tremendous secrets of the human frame', one species of fake accounts has a Frankensteinian character. They are generated by programs that snatch pictures and information from the profiles of real people, and then stitch these parts together to make a passable wretch with upwards-pointing thumbs. Victor Frankenstein displays a similar moral indifference to his sources: 'The dissecting room and the slaughter-house furnished many of my materials.' Thus, in online entourages, real people rub shoulders with these resurrected, composite monsters.

Intertwitter and its many competitors such as Fast Followerz are part of an industry of synthesised support estimated to be worth hundreds of millions of pounds. For some business owners, buying fake followers is as practical as a busker throwing a five-pound note into the hat, just to get things rolling. This online Boosterism is worryingly inexpensive. Fast Followerz will unleash a batch of 10,000 Twitter Followers onto your profile for $99, while Intertwitter offers the same for $65. Differences in price can arise from the pedigree of these fake followers themselves. Some zombies are more convincing than others in their liveliness. The most lumpen of these programmed 'bots' are easily detectable by their lack of photographs and biographical data, the infrequency of their engagement in activities such as 'liking' and retweeting, and the ratio of their followers to those following them. The internet's walking undead betray themselves by being, as in the movies, the follower more than the followed.

Since the host platform, such as Instagram, will purge accounts it deems to be inactive or fake, this industry has adapted by inventing the 'click farm'. These outfits consist of banks of computers, at which low-paid workers in countries such as Bangladesh sit in shifts.

They log in and out of thousands of zombie Facebook profiles, for instance, and 'like' the product that has paid for this boost. Since real people rather than bots are at the helm, these endorsements are more difficult to distinguish from authentic consumer behaviour. Thus we arrive once again at the horrors of uniform emotion, of humans contractually reduced for hours at a time to a fixed grin of cyber-friendship, their worth measured in harnessing an army of the undead in a grimly impressive layering of exploitation. Choosing a product based on its social-media popularity is thus now embroiled in the ethics of fair trade. By responding favourably to this currency we reinforce its value, and in turn encourage the rationale for click farms. Whereas the trademark unethical choice used to be favouring cheap coffee at the expense of underpaid labourers, in the case of counterfeit endorsements the unethical choice is to favour the goods that are rich in approval.

It is a sign of the progression of capitalism in my lifetime that as a child I was taught to be outraged by the sweat pouring from the exploited, and now thirty years later it is not just their sweat but their indiscriminate enthusiasm for the world that testifies to their oppression. I remember noticing as a teenager how the McDonald's menu hanging behind the standard tableau of disaffected servers listed smiles among the products, with the price set to 'Free'. As is typical, this giveaway means nothing to *The Man*, especially when he happens to be a clown, while presumably for his wage-labourers this disposable happiness is more costly. It would have been difficult then to predict that, in the new millennium, the uncanny Clown would be asking for our free smiles, or that when smiles did appear on the menus of 'social-media marketing services', they would have both an ethical and a monetary price attached.

Sense of an Ending

If digitisation alters the experience of life, then death can't expect
to remain untouched. Online we move among zombies, and in the
real world friends can feel like after-images. We are growing
accustomed to looking over our shoulders.

Here is one last digital ghost story. Several years ago, one of
my Parisian acquaintances was a tall, lanky pianist whose hands,
once unfurled, seemed to stretch from one end of the keyboard
to the other. He could be shown a song on YouTube and reproduce
it a few minutes later. He lived in a tiny one-room apartment near
the Asian Quarter in Belleville, and would sit at his electric piano
with his headphones on, the window open to the courtyard below,
practising a silent and difficult concerto. He drank excessively and
regularly, and if you met up with him early in the morning you
could catch him pressing his thumb and forefinger over the bridge
of his nose and closing his eyes. At home he made artless coffee
from a tub of Nescafé. When a model scout approached him on
the metro he ripped up the man's card in his face. On the bookshelf
by his front door was a disorderly collection of horror movies,
such as *Chucky: Jeu d'enfant* or *Chucky: La poupée de sang*. He was
gentle and indulgent, though owing to glitches in my fluency I
wasn't always sure what he was talking about. He had what I
noticed to be a very charming French quality, which was that if
he was with you he would be uninterested in the world beyond,
and would blow off upcoming plans to extend the rendezvous at
hand. The flipside of this immersion in the present was that he
could fall out of contact for weeks at a time.

When I left Paris we didn't keep in touch, but I would occasionally think of him filling his apartment with silent notes. One night Facebook told me that it was his thirty-third birthday. I was in bed and about to go to sleep, but decided on one last lap around the circuit of email and social media. Clicking on the birthday icon, I recognised the names of some of the friends who had left him messages. My last knowledge of him was from some so-called Facebook stalking undertaken three years previously, and he seemed to have left Paris for a provincial town. I remember not being surprised that he hadn't stayed in that tiny studio in Belleville, but I also knew that a broad circle of friends, many of them fellow musicians, would miss him. He seemed to be at the centre of a communal feeling that was a mix of protectiveness and awe. Among a few generic birthday messages, with smiley faces and hearts, were more plaintive sentiments. I felt the warmth and steadiness of a confirmed prejudice: Parisians were a melodramatic people. I hadn't read French in a while and I was rusty, but I kept hearing this dolorous note struck. You have left such a hole, one of them said; another admitted to having dreamt of him again. His lover at the time I knew him had posted some forlorn words, and so I assumed that they were freshly broken up.

Elsewhere his Facebook page was vivacious, framed with companionship and his movie interests and pictures of him clowning around. Then I read a message from an anglophone friend, who was presumably not naturally prone to Parisian melo-drama, wishing him a happy birthday 'wherever he was'. My breath began to quicken and I sat up on my elbow. I had suddenly real-ised that, among all these well-wishes and laments, my friend himself was nowhere to be found. But what name can we give to this excitement, this instant zeal that I felt? The night grew later; the pizza shop had long since shut off the fan that ground outside my window. I followed the scent of grief, back down through the posts. Shock isn't a condition that encourages our finer traits. In my case it foregrounded the sorrow that was to come later with an almost pleasurable curiosity at the puzzle before me. While the older posts were loading I put his name into the mini Google search field and saw that Google presumptuously added the word 'mort', making him hideously double-barrelled and stopping my heart for a moment. The search yielded no results.

Jonathan Safran Foer's novel *Extremely Loud and Incredibly Close* includes a run of freeze-frame images of 9/11's infamous 'Falling Man'. They are arranged in reverse chronological order, so that when you flip through the pages the man is sucked skyward, heading back into one of the upper Twin Tower windows. In the logic of this rewound world, the planes will un-fly into the buildings; the victims will un-walk themselves from the disaster. Social media's ability to bring you back into the past shadows our instinct to deny tragedy and to retreat into a safe time when the horrific thing that has happened needn't necessarily have happened at all. And so it is the summer of 2010, and the birthday boy reawakens, sanguine and full of the future. No one is sad for him. His 'activity' must negotiate with my morbid nosiness, which is the beginning, and in some cases the end, of empathy. He is planning to move back to Paris that autumn, from what I can tell, to enrol in a musical conservatory; there are posts of him playing the piano and singing songs in small bars and coffee shops. Occasionally he writes teasing messages to friends. Now that he is re-alive, I find that I need to locate the precise day when all of that stops, all over again. It feels like homing in on a crime scene, and all the while there is that excitement that seems at once to have everything and nothing to do with being human.

His last few posts were music videos: the Ronettes, Broadcast's 'Come On Let's Go', a nostalgic nod to Natalie Imbruglia. And then it occurred, manifesting on his relatively sparse timeline like a pile-up of traffic: friends keening at him, or through him at each other; his father possessing his mother's account to gather information and make almost stern, efficient arrangements. I recognised a female friend of his, whom I'd met once, among the anguished voices. She had appeared in one of his music videos with live snails placed here and there on her body. It came as a relief to find this scalded patch of his profile. He had fallen away as though through a trapdoor, and suddenly he was colonised and animated by others. What an appetite I had for their posts, and there was little else besides garden-variety disappointment when they began, up through the days and weeks, to thin away. It was unclear what had happened to him, and to this day Facebook respects this mystery.

Next I clicked across his pictures, wanting to look him in the eyes as best I could. We were born in the same year, and I felt a mammal's wonder that he had managed to grow so cold. But then again, there he is, smoking and wanly smiling. One afternoon in Paris we were standing on a bus. I pointed out an old woman playing sudoku, though now I can't remember why. 'I feel sad when I see that,' he said in English, gently banging his shoulder against the pole.

I had to be up early and my eyes were getting tired, but I was now hooked to the aftermath. In the quiet of that first autumn, when friends came less to the cyber-gravesite, it was of course the mother who kept returning, standing in the bleak weather of his silence. She called out his name, asking him to come back. She spoke to him about going to the Leonard Cohen concert for which he had bought her tickets, and how in the songs they were reunited. Facebook was no doubt one of several places where she went to commune with him, but nevertheless she must have felt that she could reach him there. Did Mark Zuckerberg imagine, during those nights of fevered coding in Palo Alto, that he was building, among other things, channels into the underworld? The mother's presence on his page seemed hostile to my gaze. The depth of her grief reproached my nocturnal curiosity, and she subdued the adrenalin that had been making me theorise edgily about *how it had happened*. These amplified graveside whispers are in themselves an intruder alarm. They scrutinise one's own bad faith, and they activate those deterrent emotions of shame and guilt. Yet even these warnings were neutered by my invisibility, by my permitted trespass, which seemed both victimless and victim-filled as I haunted a space some-where between gossip and literature and memory.

After a while, the mother's beseeching stopped too. I wended my way back towards where I had started, to the mournful birthday party of the present. Time began moving quicker as the posts became more sporadic. But the mother was on my mind, and so I clicked on her profile, summoning her up now to answer to me. My next witness. Near the top of her profile were a few friends bidding her farewell. I realised that she had died almost exactly one year after her son, and that, in the depths of the night, I had been listening to one ghost mourn for another.

From my open bedroom door, I could see out into the darkened kitchen. The saucepans were hanging in the shadows, as expected, and the road beyond was quiet. I closed my laptop and lowered it to the ground. The power cord's overheated box knocked against the floorboard. I leaned over the side of the bed to set my phone's alarm. Then I clicked off my bedside light, and tried to settle into sleep.

6

The Cabin in the Woods

I f sleep does come, what then? Of what does a four-dimensional human dream? For sleep, like everything else, is being digitised. A study has recently suggested that our sleep hormones are being blanched from the light of our various night-time screens. As a result we may sleep less well, and wake up more tired. Our phones are black on the bedside table, sleeping the sleep of the just, but their light swirls on behind our closed eyes, so the survey says. We resolve, in the dark, to get an old-school alarm clock, and to keep the phones and laptops on the other side of the bedroom door. Everything must go. Except there's no landline now, sitting venerably on that little table by the stairs. What if someone needs to call us in the middle of the night? That is a call we should take. And so the phone stays resting on the nightstand, undisturbed.

Oh, those studies. Despite them being everywhere, most of us, most of the time, see them out of the corners of our eyes; we catch them fluttering down the pavements, we hear snatches of them on the wind. Most of us have little inclination to spend much time at the coal face of raw findings. But we get the gist, and the gist is not good. As well as stunning our melatonin like a deer in the headlights, our screens are, it seems, less easy to learn from. While there is still much debate, there have been enough proclamations for these headlines to bypass our brains and go straight into our nervous systems. We worry for our concentrations, our ability to absorb information and to memorise. It is easy and understandable to feel that we are running our own labs, producing reams of intuition, if not data. A friend's mother warned him not to keep his phone in his pocket – 'you're microwaving your testicles'. No body part, or faculty, is safe. Meanwhile, studies about the health benefits of pet ownership glow with good tidings, year on year.

Fearing the toxicity of our gadgets is not a new pastime. On Christmas Day in 1991, my aunt, then in her late fifties, sat in her dressing gown with the electronic memory game Simon in her lap. On my request she was giving this new present a reluctant go, tapping at its four big, coloured buttons as they lit up in sequence. I remember Simon flashing and bleeping with the ponderous, patronising lethargy of its easiest setting. When it eventually caught her out she passed the plump little flying saucer back to me and straightened the skirts of her dressing gown. 'They say all this electrical stuff will give us cancer,' she remarked. Similarly, like every post-1950s generation, when I was young I was told not to sit too closely to the television. But while the pastime isn't new, the stakes have been raised by our latter-day habit of fondling such gadgets from the moment we wake to the clunk of the power cord and the click of the light. Simon, by contrast, knew his place. Now I take the television into bed with me, and it lies warmly, companionably against my stomach. A degree of lassitude usually accompanies this pose, but nevertheless I entertain an occasional thought for my organs beneath that spreading warmth, and I wonder how they're taking to all of this, what grudge is accumulating night by night, and what they have in store.

The studies, alas, don't stop with toxicity. Another narrative running alongside digital progress is one of emotional fragility, of depression and addiction. Science is regularly finding new ways that social media is making people sadder, though whether it is generating rather than displacing sadness is difficult to judge. Regardless, the effect of this pervasive suspicion can be as enfeebling as someone repeatedly telling you how tired you look. Anecdotally, this discontent seems apparent enough. You know that the party is definitely over when conversation becomes a healing circle for Facebook sufferers. In the last year, my students have begun to mention the guilt of what they call 'Netflix binges', suggesting that, with their mouths ever open to a menu of online 'feeds', they're imagining themselves as gluttons. Unsurprisingly, then, a Lenten spirit is emerging as a counterbalance. Internet usage (the social-media wing in particular) is now a staple denial on Lent's annual blacklist. The restaurant game 'Phone Stack' acknowledges our compulsions: in the game the diners pile their phones in the middle of the table and the first person to make a grab during the meal picks up the bill.

But from where does this feeling of taint arise, and how does it come to run in perfect parallel with our online enthusiasms and pleasures? When social media works, it feels so like the everyday conviviality and friendship of physical interactions that one could nearly forget to give it any credit. 'Social media' as a formal topic is generally a troubled one. At such times it is characterised as an addictive substance, a depressant that needs to be managed and curbed. Everyone knows someone perpetually on the brink of quitting Facebook. One hypothesis for this malaise suggests that the gloss people give to their crafted online personae creates an epidemic of inferiority among those of us watching them, as well as an ongoing amnesia towards our own canny uploads. This gulf widens when you consider where we do this watching: from the unglamorous heap of our insomnia, on the choked bus to work, the windows so fogged with kettled breath that you can't even dream out of them. If this hypothesis is true, then when people log on to social media, they are apt to feel as though entering a joust, with their friends' successes and wit, their general robustness for life, coming at them like lances. The overt and odious phrase 'You win the internet' – deployed to congratulate an exemplary piece of digital behaviour – does nothing to soothe suspicions of tacit competitiveness.

The inferiority theory is too easy to be the whole story, but nevertheless there is a strong and widespread feeling that our relationship with digital technologies has to be managed as a sort of chronic problem. Simultaneously we are rightly enamoured with all the ease and enrichment they provide. The four-dimensional human thus regularly experiences two types of breathlessness. The first is due to the thrill of roving over the world, of dropping in on a sibling and their baby on another continent, of staying for five minutes and laughing the whole time, then swooping back into your skin. The second breathlessness is not cheerful, and arises in the moments when all this liberty seems to come at the price of its opposite, when the sum of digital life feels more like a cage than a flying carpet. The ongoing narrative of toxicity and depression that shadows digital progress, in conjunction with a sense that this progress is both for the best and inevitable, creates a pervasive atmosphere of claustrophobia. The weather is often close in the fourth dimension. A small signal of this confinement is how the phrase 'surfing the internet' is used much less frequently now than in the old days. Rather than coursing the waves, we are simply, immovably, *online*. When multiple aspects of digital life are consistently figured as sources, suspected or confirmed, of bodily and psychological pollution, then our irreversible journey deeper and deeper into the network can, in one's less hardy moments, feel like an imprisonment. As a result, when sleep does come, the four-dimensional human begins to dream of escape.

Claustrophobia descends swiftly in moments of thwarted movement, which digital life provides in abundance. Much of online ennui arises from this sense of being on the wrong side of a locked door. We can spend hours rattling the handle. Sometimes the craving for motion, almost as an abstract idea, movement for its own sake, leads to a temporary, claustrophobic madness. For instance, have you ever scrolled the drop-down menu of your favourite URLs and selected a website (mentally licking your lips and settling yourself more comfortably), only to realise that you're already on the site itself? Here our destination has become our beginning. When the website helplessly judders and copies itself, it seems to ask: What do you want from me? Where are you trying to get to?

The ironic, unsparing conclusion to Aaron Sorkin's Facebook biopic, *The Social Network*, shows a classic mode of this confining online behaviour. A fictional Mark Zuckerberg sits in a deserted conference room, as rich as a king but lonesome. With an expression that blends fatigue with fear, he repeatedly, fanatically, clicks refresh on his browser, to see if the pretty woman has accepted the Friend Request that he has sent her just seconds before. This is dramatically a full circle of sorts, since this woman dumps him in the film's opening scene. And yet, the ending is purposefully a non-ending, and instead Zuckerberg experiences, not romantic 'closure', but the breathlessness of a closed loop of his own design, in which he alone is trapped. Here, the conclusive move of 'boy gets girl back' is perpetually delayed, and each cycle of refreshment leads to the same place. You don't have to be lovelorn to experience this parching type of refreshment (but it helps).

There are many varieties of the Zuckerberg Loop or, if abbreviated like a weblog, the gloop. Running laps around one's track of message-bearing websites, from most to least exciting – work email, personal email, social media – is a classic gloop, as is the insistent return to ongoing news stories or sports scores. Our weakness for these motions without progress feeds the argument of digital life's addictive properties. It is as though the early internet's dream of freedom, in its most distilled form, is now parcelled to us as the habit-forming gaps between loaded pages. This gap is a kind of bodiless free fall, a rush of pure potential in which life feels up for grabs. As internet speeds increase, it becomes less evident how hooked we are on this infinitesimal pause, a moment of wild possibility, which can so consistently resemble that false plunge at the verge of sleep. My long-serving, virus-riddled computer is annoyingly sluggish, but its dilatory transitions exaggerate and reveal the enclosed end to this jittery feeling of suspense and optimism, like the looped, slow-motion footage of a Venus flytrap snapping shut. A major source of digital claustrophobia is the culmination of a thousand of these tiny failed escapes.

To me, these gaps contain a beguiling pastoral romance. They seem to be the place where the peace of the countryside, the freedom of a steam train winding through a valley, can be achieved, except that it can only be experienced as a series of bursts, like glorious puffs of a cigarette. This is a journey made in interrupted inches. And much like smoking, this practice involves two ends of a spectrum brought together in a paradoxical loop, creating the confining release, the heart-pounding relaxation. Especially when I get trapped in that airless circuit of websites, a journey to where I began, with nothing changing, no message, no sign from beyond, no betterment, it is easy to feel the little neurological paradox of a motion that goes nowhere, more familiarly known as the shudder.

While we can bring this feeling of confinement on ourselves, there are often other people involved. Being coerced into doing something is another shortcut to claustrophobia. I never mind social media as much as when someone tells me that we all need to embrace it because 'it's here to stay'. I'm not always sure that being 'friends' with my local curry house is the way I would like the world to go. Start-up businesses, young interns often tell me, 'have no choice' when it comes to participating in the digital status quo. Enterprise must promote itself in the big online arenas, and many would argue that this prospect is as alarming as having to hang a sign above one's shop. But digital life has the habit of entwining commercial and personal concerns, and the same young interns also say that social life would be impossible without digital connectivity; their recent adolescence was, whether they liked it or not, locked into its rhythms and opportunities. It is hard not to feel a certain amount of duress on hearing such things.

The logic behind this pressure has the circularity of an arms race – everyone believes it is necessary, so therefore it is necessary for everyone. The ability to engage in these new technologies is generally seen as an essential aspect of citizenship. For the uninitiated, learning the ways of digital life is presented as a form of salvation, while in parallel the studies keep coming, and exhausted millennials run their hands through their hair and voice the wish that they could 'turn off the internet'. The numbers of uninitiated are certainly dwindling, and in these times there is the sense that they are being benevolently rounded up. Come with me!

In this vein, Barclays bank has recently released a flock of 12,000 'Digital Eagles', who offer free tutorials to those lacking cyber-literacy, 'even if you don't bank with us'. The elderly are obvious beneficiaries of this programme, though their lives in the new millennium are often presented as a choice between crushing lone-liness and online connectivity. The chance for corporations to take bites out of our sociability, as compared for example to an unof-ficial network of concerned neighbours, is steadily turning commu-nity itself into a commodity. However, in one of the Digital Eagle adverts, it is a middle-aged Welsh miner who is floundering in cyberspace. Community here isn't the problem. The scene is the antique valley-scape of a 'tight knit' small town in Wales. Digital progress is shown as both a scourge and a fulfilment – 'It is taking over the world, to be honest,' says the Digital Eagle with a tone of innocent concern, while the shy miner emerges dolefully from his shaft, the polar opposite of a virtual office. The skills he learns put a smile on his face and an email in his inbox: 'Welcome to the twenty-first century, Dad!' Here is the temporary terminus of the gloop (more of a lay-by), and the miner is, for the moment, at ease in his new home. Smeared in coal, he is the one who seems to have been excavated, as though it is Barclays who is mining the dark places of the earth, dragging up the last souls never to have seen the light of a computer screen.

Gridlock

Few events put constraints on one's hopes and dreams like the end of the world, and in 2012, the year of the Mayan prophecy, doomsday was definitely a thing. The worst quality of apocalypse is the lack of choice it offers. Once it finally arrives, it is as inevitable as a start-up business's Twitter campaign. But while most people laughed off – more or less nervously – this spectacular prediction, the prophecy suited the claustrophobia in the humid cultural air. It was also in this year that three big-budget, speculative-fiction films found metaphors for our present confinements. While the uninitiated were being reminded that digital life was here to stay, these films each articulated the oppressiveness of existing involuntarily inside a scrutinised network, and perhaps what makes their respective worlds feel most futuristic is the intensity of their nostalgia for an unwired life. These films show people suffering from being, in various ways, confined, and their strategies for escaping the grid into which everyone is expected to be hooked. They feel forced to move inside too little space, and are prey to those who watch them. From these conditions various fantasies of wilderness arise, pastoral romances for caged creatures.

Catwoman in Christopher Nolan's *The Dark Knight Rises* isn't purring over a diamond or the prospect of seizing power and riches, but rather the escape promised by a computer program called Clean Slate, which wipes a criminal from all global databases. Absence and anonymity are the new treasures. Played by Anne Hathaway, this Catwoman has a Normcore fantasy of evaporation, free from the confinements of her digitised history, inside which she is criminally locked. She wants to run in an unmapped place, to steal the right to be forgotten. In the 1992 Michelle Pfeiffer rendition, by contrast, Catwoman is vengeful and involved. She wants *in* rather than out, since all the action and the glory is inside the system. At the film's conclusion, although she is assumed to be dead, she doesn't take her chance to flee. Instead, her shadow can be seen slinking away down alleys, evasive but present, refusing to be lost. The final shot shows her looking up at the powdery cone of the Bat signal, defiantly still in the game, somewhere in the city. Her later incarnation, however, dreams of a wilderness beyond the city, where she might embrace a soothing blankness.

The gangsters in Rian Johnson's futuristic movie *Looper* share millennial Catwoman's fatigue with traceability and her thirst for the open air. In this time-travel film, a young man from the year 2044 is paid to carry out executions for criminals living thirty years in the future. This outsourcing is necessary because, by 2074, surveillance and tracking technologies have made it almost impossible to dispose of a stiff. No murky canal or convenient thicket now sits unwatched. So, using cutting-edge time machines, these mob bosses send their victims back to the quaint 2040s, where someone (or their corpse) can still fall from the grid. The film's central drama, then, revolves around the intense claustrophobia of the future, which manifests in various forms. Executioners such as Joe are eventually sent their future selves to dispose of, thus 'closing the loop'. When old Joe escapes the one-man firing squad of his younger self, an implosive chase begins in which the hunted and the hunter are the same person, ending up in the heroine's remote farmstead. There's no longer room in this futuristic gangland for the goodie and the baddie to be different people. Even the criminal world's fashion is caught in a pathetic cycle: young Joe's boss, an aging gangster who has relocated permanently to his own past to coordinate these hired guns, teases Joe for his 1950s-esque pomaded, leather-jacketed, *hood* style: 'The movies that you're dressing like are just copying other movies. It's goddamn twentieth-century affectations. Do something new, huh? Put a glowing thing around your neck, or use rubberised . . . just be new.' Here again we see glimpses of how a hyper-connected environment breeds a crisis of originality. A culture of sartorial recycling is, according to *Looper*, the dress rehearsal to a standing-room-only world.

In the third of this claustrophobic triptych, Drew Goddard's *The Cabin in the Woods*, five American college students pile into a camper van and attempt a weekend in a remote getaway. They're excited by the freedom of the road and their desolate destination. 'It doesn't even show up on the GPS,' one of them says, holding up a stumped smartphone. 'It's unworthy of global positioning.' To which the drawling stoner replies: 'That's the whole point, get off the grid, right?' In one version of the film's trailer, the captions invite you to remember the well-ploughed territory of this horror scenario: 'You think you know the story . . . You think you know the place . . .' These taunts refer to the twist that, in trying to escape the mapped world, these college kids have unknowingly journeyed into the centre of surveillance. For you see, this cabin in the wilderness is actually the monitored and highly controlled platform for a secret, ritual blood offering to subterranean Ancient Ones, an annual sacrifice coordinated by a specialised facility that stage-manages every aspect of the horror. Teams of workers in this high-tech institute chat mildly about their weekends while making sure that the killing goes smoothly, that the 'holidaying' kids get drunk and horny and careless on cue, that the zombies are unleashed according to protocol. Thus we find a version, on the grand scale, of failed refreshment, the thwarted getaway of the obstinate website, where people journey and go nowhere.

Cabin is the next stage of the postmodern knowingness on display in *Scream*. In *Scream* ironic self-consciousness – everyone in the horror movie discussing horror-movie tropes – is one of the pleasures of Wes Craven's engagement with genre, but, despite all the winking, ultimately the middle-of-nowhere house remains true to its word: the remoteness and the black of the night are as sincere as the sharpness of the ghost-faced killers' blades. The ironic loop that has a babysitter on the brink of being murdered chatting about murdered babysitters is itself a claustrophobic aesthetic, but the vice is further tightened in *Cabin* by the fact that the haunted house is a simulation, its middle-of-nowhere quality merely an affect. The eponymous, four-dimensional cabin, as the '*The*' in the title suggests, is an archetype of lonesomeness. In the current lingo, the film could equally be called *That Cabin*. Moreover, the young victims are explicitly chosen as archetypal horror-movie characters: the whore, the athlete, the virgin, the scholar, the fool. Whereas *Scream*'s irony exposes us, the viewers, by comically baiting our own literacy of stock horror, in *Cabin* the exhaustion of the slasher film is intensified by the presence of another audience between us and the bloodshed. It seems we can't even be slaughtered in peace these days, without somebody watching.

All Sewn Up

It is companionable but not uplifting that the personal sensations of digital claustrophobia are in keeping with our era's broader economic and political concerns. The very notion of contemporary citizenship has been compressed, which is reason enough for breathlessness. One of the consequences of having consumerism at the centre of civic life is that the market becomes a primary site of ethical expression. If the Greek agora was a place of both politics and commerce, then we have a crowded two-in-one deal: speechifying and shopping combined into politicised consumption. With this Head and Shoulders amalgam, we make an argument with every purchase. The feeling of entrapment arises in the fact that we may unwittingly be debating for the Nays who are packaged as Ayes, because in the gloom of late capitalism, it can be hard to tell who is who. As Robert Schumann wrote in his creeping, crepuscular song in the *Eichendorff Liederkreis* cycle, which tells of shadows emerging from the woods, 'If you have a friend on this earth, / Don't trust him at this hour, / Friendly perhaps in glance and voice, / He's planning war in deceptive peace'. A globalised marketplace means that the ethical consumer must navigate a densely interwoven forest of corporate family trees. The challenge is not simply to have the privilege and decency to select the saintly product, but not to be taken in by false friends. It is easy to feel that there is no way out of unethical decision-making.

Our modern economic claustrophobia is triggered by the mere mention of the corporate giant Monsanto, against which many people have huge ethical objections, while also feeling that there is no consumer-political space in which to launch their counter-attack. Monsanto has become a well-known threat to an unpolluted pastoral. Its plot, as circulated by activist rebels, sounds too diabolic to be true: 'I will own all the seeds,' the giant says, 'all the seeds in the entire world will be mine, and I own all the factories, too, not just seed factories but every sort, down to the last one.' The spectre of Monsanto haunts the supermarkets, and is no doubt made more fearful by paralytic ignorance, by only knowing its name through apocalyptic rumour and by superstitiously signing online petitions. You suspect that you can't walk into the street without inadvertently lining the ogre's pockets. In some hideous butterfly effect, with each dip into your mustard pot a swarm of bees falls choking from the cloudless Idaho skies. To be good in this world, you need to send every grocery item onto *Who Do You Think You Are?* The supermarket shelves are an identity parade of their own, with the moral shopper forced into forensic detection, asking Mr Clean for a sample and inspecting Aunt Jemima's headscarf for a fibre of Nestlé. We get home and hope there isn't blood in the Campbell's. We're looking for a clear lineage, a virtuous pedigree, which runs all the way back to that unique, organic smallholding, where the animals lead fulfilling lives (to a point) and where the well-paid labourers dine alfresco, clinking their glasses of Chianti as twilight comes folding in from the orchard. The shopping baskets of the righteous brim with pure stock. Corporate intermixing is unwelcome because, unlike the genetic variety, it increases rather than reduces the threat of pathological transmission. Money has no immune system.

It was appropriate to the spirit our age that, while I sat darkly mulling the binds of neo-liberalism, a get-rich-quick scheme occurred to me. Similarly in keeping was the digital nature of my capitalist venture. I would design an app – my own fiscal escape route – that *was* a shopping version of *Who Do You Think You*

Are? It would be called, perhaps, Supermarket Sweep, and would enable consumers, from a simple phone-scan of a barcode, to trace a product's corporate lineage. In this way, you could instantly determine the ethical pedigree of everything you buy, and thus purchase, at long last, an uneventful night's sleep. I savoured the promise of riches that would soon accumulate in millions of tiny increments, as relieved shoppers imported my idea, for a few honest pennies, into their phones. My own wealth would be pure as tundra. All I needed was a database of corporate family trees, so in the name of research I put a few keywords into Google. The results were in a private way confining, since the following entry appeared to *Thelma & Louise* me all over again:

> Have you ever wondered whether the money you spend ends up funding causes you oppose?
>
> When you use Buycott to scan a product, it will look up the product, determine what brand it belongs to, and figure out what company owns that brand (and who owns that company, ad infinitum). It will then cross-check the product owners against the companies and brands included in the campaigns you've joined, in order to tell you if the scanned product conflicts with one of your campaign commitments.

The sitcom *30 Rock* parodies the economic claustrophobia from which the Buycott app arose when Liz Lemon buys the perfect pair of jeans from what she thinks is a hip, independent shop called Brooklyn Without Limits. In commercial terms, wilderness lives in the moral haven of the independent shop; mercantile authenticity offers a kind of store-bought pastoral. So, feeling both sexy and economically pious, Liz flaunts her bottom and her social conscience, proudly displaying the 'Handmade in USA' label stitched on the back. But then her Big Business boss, Jack Donaghy, reveals the provenance of Brooklyn Without Limits: it was not, as its own myth-making goes, founded by cool, anti-corporate Brooklyn Zack, but rather by the multinational, multifaceted oil corporation Halliburton. As for its honest, home-grown label: 'The Hand people are a Vietnamese slave tribe and Usa is their island prison'. When Liz protests that she would have read about this scandal in the liberal media, Jack replies that there's no such thing, reeling off a fictive genealogy of corporate subsidiaries: 'The *New York Times* is owned by NYT Incorporated which is owned by Altheon Ballistic Dynamics which is owned by the Murdoch family who are owned . . . by Halliburton.' There is a moment of jokey Gothic fright at this revelation: Liz gasps as the music goes Germanic. In Brooklyn Without Limits, the mannequins wear straightjackets, which Liz's friend explains is a sly reference to the shop's site being a former psych ward. But it's the consumers whose arms are bound: in choosing the trappings of 'independence', they accidentally endorse the dubious ethics of the multinational. Alternatives turn out to be false choices, and escape routes loop back to the cell. Indeed this is one of the Gothic's tricks: to give the illusion of escape, when in fact you've never actually left the confines of the spaces that haunt you.

I Would Prefer Not To

It doesn't seem accidental that a major global resistance movement to emerge in these claustrophobic times has been Occupy. The problem of the one per cent controlling too much of the world's wealth is, among other things, a problem of blocked motion: in this case, the circulation of capital. An outcome of this inequality is that money, which was invented to flow, is trapped inside vaults. But more than this, Occupy represents our confinement in the mode of its resistance. In a world stripped of the possibility of wild places, the only option is to trespass. In New York in 2011, Occupiers in Zuccotti Park exposed the illusion of public space when then-mayor Michael Bloomberg had them evicted from their encampment. A literary figurehead of Occupy is the subject of Herman Melville's 1853 short story 'Bartleby, the Scrivener: A Story of Wall Street'. Bartleby is hired as a clerk in a legal firm and gradually rejects the tasks assigned by his relatively benign employer, replying placidly to these requests with an unwavering phrase that arrestingly combines yes and no into a positive negation. 'I would prefer not to.' Bartleby's relationship to Occupy becomes apparent when we learn that he is sleeping overnight in the law offices and that, when his boss's patience runs out, he ignores all demands to vacate. The philosopher Slavoj Žižek found Occupy to be an example of a 'Bartleby' style of resistance, one which stalls the status quo's operations and established power structures while offering no coherent alternatives.

A frequent initial critique of Occupy was its lack of positive demands, which seemed a strange response to those aware of Occupy's explicit insistence on the separation of corporate interests and democratic process. Nevertheless, the strangeness of a movement without progressive motion was an essential aspect of the resistance. Critics argued that a demand in itself legitimises the authority of the very institutions that Occupy seemingly was seeking to oppose. Here again is the breathlessness of being caught by the throat: how can you voice a plan of liberation when the language available, all the vocabulary, is restricted to the narrow, self-perpetuating terms of those from whom you seek freedom? Bartleby, the copyist who refuses to copy, sits amid the blank pages of undone work, but he does not use them to write a *Jerry Maguire* manifesto for change. Similarly, Žižek finds the inarticulacy of Occupy to be symbolic of its opposition to a capitalist democratic system that absorbs all resistance within its own ideology. Years before Occupy, Žižek noted how the 2003 protests against the Iraq War suited interventionist governments by being in themselves proof of the supposed libertarianism of western democracies, the political ethos that the invasion claimed to be defending.

That both agreement and opposition, Aye and Nay, can be put to the use of those in power reveals another collapsed polarity. This collapse of meaning is one of the most important aspects of Melville's story. The grotesquely static Bartleby, spending hours in trance-like states, embodies the petrifying qualities of authoritarian power structures that uphold corporate architecture. His employer tries to indulge Bartleby's refusal, and encourages him to take 'wholesome exercise in the open air. This, however, he did not do.' Melville is telling us that in this enclosed environment of screens and walls there is no such thing as open air. Bartleby is ultimately arrested and removed for being a vagrant, but the irony of this indictment reflects the nonsensical extremes to which language, in such a system, is pushed. His employer recognises the lunacy of this charge: 'What! he a vagrant, a wanderer, who refuses to budge? It is because he will *not* be a vagrant, then, that you seek to count him *as* a vagrant. That is too absurd.'

Melville was quick to sense the confinements of modern professional hierarchies and the paradoxes that their monoculture produces. Bartleby, as the static-vagrant, is a monstrous impossibility. Besides being the patron saint of Occupy, in the profundity of his ennui he is a prophetic anticipation of Aaron Sorkin's final image of Zuckerberg. The clerk spends hours entranced by the view out of his window, which is the unchanging non-view of a wall, the essence of failed refreshment. Indeed, Zuckerberg's repeated refreshing of his browser is an echo of Bartleby's appointed task of copying out documents. Both of them are caught in a cycle of reproducing that which already exists. Being absorbed into this strange form of static work, motion without progress, is a hazard of online life.

The Hermit's Hut

Where do we look, now, for the open air? It comes in strange gusts these days. Recently I wandered by chance into a more remote corner of the internet, where there sits a web page. It is a long-forgotten and forgettable *Guardian* article from 2003, but I'll say no more about it. To name it would be to rob it of its retiring allure. I only remember it because of an unusual set of architectural features. In the top right corner, it displays the now-standard moulding of social-media icons, but since this minor article dates from late Web 1.0, it unsurprisingly has no social-media presence. What might we call this mock-modernisation? It's the opposite of putting in a non-working Victorian fireplace, and is more like adding ornamental solar panels to a thatched roof. For alongside each of those futuristic tallies – Facebook Shares, Tweets, G+1 and 'in Share' (whatever that is) – there was a defiant nought.

That string of zeroes, as cool as digital midnight, made an impression on me. It gave the article an unintentional aloofness, the sort that young children can have when they look out at you from black-and-white photographs. It might be simplest to say that I felt glad for it.

My response to this web page is both neurotic and true. I should note that there's a charmless asymmetry to this reaction, in comparison to the bawdy glee with which I've monitored the numbers rise on my own virtual offerings. In such instances I want each of those sacks stuffed to bursting with digits, bulging with that private, mostly insolvent currency whose coins are stamped with my own face. But overall I can track a change in myself through my response to the online zero. In my mid-twenties, before the cloak of invisibility years, I would ride the rapids of the Facebook chat function. Despite the sexual comedy of errors locked into its coding – the brutal curtness or the hot-cheeked graciousness, depending on which side of a failed gambit you find yourself on – I could be prone to anomie on discovering, at certain odd hours, that (o) Friends were online. That Munchian zero triggered a half-joking angst, which could simultaneously be felt and laughed away. There was no angst, however, having hit the comparative churlishness of my thirties, in discovering that unshared *Guardian* article and its sheltered solitude. This website in the woods had the aura of having been spared something; those four zeroes, detached links of a chain, contained in each cold loop a primal sense of fastness and calm.

In *The Poetics of Space*, a study of our mysterious intimacies with the rooms in which we live, Gaston Bachelard writes of the 'hut dream'. Here he traces the experience of wonder and yearning we're apt to feel when we come upon, or even imagine, a lone dwelling tucked safely within its anonymity. 'When we are lost in darkness and see a distant glimmer of light,' Bachelard writes, 'who does not dream of a thatched cottage or, to go more deeply still into legend, of a hermit's hut?'

Who, indeed? Is there anyone out there?

Years ago, on a midsummer evening, I went cycling with my older brother and his work friend through the rural lanes of one of the London commuter shires, stopping at country pubs along the way. I wasn't living in England at the time, and so the experience was like, if such a thing existed, a heroin shot of Englishness: strong ales with names like Flowers did the mellowing, and the night fell slowly over the meadows with a non-North American languor. I may have always remembered this picturesque evening for these reasons alone, but one moment pins memory to it and it is this moment that I think of first, with the rest of the scene radiating from it. 'Jean Marsh lives there,' my brother's friend said, pointing over the hedgerows, but because we were cycling downhill and he wasn't talking directly to me and the stars were by now coming out of the deep, we were past Jean Marsh's house before I could find it on the darkling horizon. Jean Marsh! Another hit of English heroin. My not-seeing ensured the moment's significance, since for ever after I have imagined the single silhouette of a house across a distance that never dwindles, a lamp in a window that borrows and duplicates the stocky twinkle of the evening star, which I may not have seen either, on that particular night.

While I can't vouch for the hermitism of Jean Marsh, nevertheless her house stands in my memory among a community of lone havens, a mental neighbourhood of vast grounds containing in the centre of each a dwelling belonging to, for example, Tom Bombadil, Crooked Finger from *Antonia's Line*, the Narnian Beavers, Jeanette

Winterson circa 1994, and Badger. 'We are hypnotised by solitude,' Bachelard suggests, 'hypnotised by the gaze of the solitary house; and the tie that binds us to it is so strong that we begin to dream of nothing but a solitary house in the night.' In this scheme, the hut represents the purest form of inhabitation; the light coming from its window is the essence of nocturnal toil and spiritual vigilance. The hermit's hut is the *Scream* house inside out: the terror of inhabiting a glasshouse at night, with a face at every window, is inverted in this image into the safety of the sheltered home in the deep wilderness, a place that keeps benevolent watch while being, by the rules of this dream, always too remote to be scrutinised itself. Bachelard writes how 'its light becomes for me, before me, a house that is looking out – its turn now! – through the keyhole'. When we imagine such places, we are simultaneously within and beyond them, too distant to plot harm and too intimate to wish it. This neither/nor, which conjures both the sovereignty and the sameness of others, opens up a poetic space for empathy.

On a well-remembered page from my earliest reading, Rumpelstiltskin cavorts outside his remote dwelling in the woods. He is convinced, understandably, that the queen will not guess his ludicrous name, and so, according to their deal, will have to relinquish her firstborn to him. His wager hinges on the robustness of anonymity, of the encrypted moniker. Since he is not living in a digital world, he feels confident enough to make up a gloating song that goes:

> Although today I brew and bake,
> Tomorrow the queen's own child I'll take.
> This guessing game she'll never win,
> For my name is Rumpelstiltskin.

My attraction to a wild place is such that whenever I sing that song to myself, the cosiness of reclusive brewing and baking entirely drowns out what I regard to be the kidnapping subplot. In other words, I remember Rumpelstiltskin for his hermit's domesticity, not his baby-snatching. The ambiguity at the heart of the wilderness romance – the cosiness and the kidnapping – is its essential feature. For the durability of such isolation is ambiguous too – the queen's messenger overhears Rumpelstiltskin singing that song, and foils the plot. But for those who like to dream of the possibility of being lost to civilisation, the fairy tale has a doubly happy ending: the queen has her baby, and as for Rumpelstiltskin, he 'stumped furiously out of the room and was never heard of again'.

Will the hut dream continue to matter, to be dreamt, in a four-dimensional world? Will its summoning up of pure, isolated habitation be a quality for which we yearn in the future? The more we come to think of a quintessential house as being lit not just by lamps at the window but by the oblong glows of assorted devices, by the half-hidden flicker of the router under the table, the less a house on a hill will be able to accommodate our dreams of solitude. Its remoteness can no longer be assumed at a distance, needing to be satisfied in four dimensions rather than three. Romance is spoiled when the distant glimmer in the trees could be equally a lonely outpost or the home base of an internet start-up. With a dimming of real-world promptings, we may find that the abstract dream itself becomes more insistent. The idea of the hermitage may become more vivid the more it becomes an impossibility, though it will no doubt take on new forms. In the digital age, quite unlike in any other modern era, we'll have to find new ways to dream of a house in the wilderness.

Going to Ground

An environment of depressive progress has nurtured companies such as Digital Detox, based in Oakland, California. Digital Detox's founders 'Are Not Luddites', the website says carefully, because to speak out against technology has become its own form of immorality, as seen in the respective beatification and demonisation of Steve Jobs and Jonathan Franzen. To give up meat for Lent is more spiritually nuanced than simply questioning meat's capacity to nourish, and likewise Digital Detox's relationship to digitisation is, on the Zuckerbergian scale, 'complicated'. However, they are unambivalent about one thing: 'The era of burnout, FOMO, multitasking, tech dependence, fatigue, "social media everything" and information overload is coming to a close.' I can't help but be reminded here of Germaine Greer's reply when people suggest that global feminism is over: 'Global feminism did not fail,' she says, 'global feminism hasn't begun!'

As part of their rehabilitation scheme, Digital Detox acquired another cabin in the woods, which promised a genuine escape from networked life. In 2013, they launched Camp Grounded in a former 1950s Boy Scouts retreat on the northern fork of the Navarro River. Among the sequoias, in 'timeless' cabins, the attendees of Camp Grounded shed their digital selves to recreate an idealised vision of American summer-camp innocence. Although its overseers are apparently well-meaning, and we could all use more yoga, meditation and good nutrition, Camp Grounded is arguably as much a simulacrum as the horror-movie creation. Instead of a killing floor, there's a 'Wellness Yurt and Tea Lounge'. The camp is unabashed about its retrograde quality as a place 'Where grown-ups go to unplug, get away and be kids again'. It promises to be 'just like the camp you remember from your childhood and favorite summertime movie', where '200+ lucky campers will take over this nostalgic scouts camp to celebrate what it means to be alive'. On a list of good-hearted pastimes (woodworking, pickling, analogue photography and archery), two in particular betray the pathos of the endeavour. The activity named 'First Love All Over Again' is an oxymoron that the illustrative beaming couple can't outshine, but most melancholic on this list is surely 'Sneaking Out' – that adolescent, nocturnal transgression. Here we have the closed loop of sanctioned disobedience, of rebellion without criminality, where interdiction and permission paradoxically join hands around the campfire.

I'm all for getting away from things and having meaningful moments with new friends, as Camp Grounded promises. But the style in which Digital Detox delivers this service, a retreat into an artificial past, both reveals and reinforces our claustrophobia. The past is the only refuge, it seems to say, because the present contains no escape routes. The implication is that only reruns of personal history, where 'chefs will be serving up healthy and delicious meals reminiscent of your favorite childhood dishes', can offer a viable rebuttal to the demands of digitisation. It is surely an indictment of modern life that, in order 'to feel alive', one must travel into the dead time of recycled adolescence. Moreover, in the case of Camp Grounded, the escape route offered is an escape into an impossibility: a childhood camp for adults.

As is common with simulations, Camp Grounded has the form of a simile with none of the novelty: your food will be *like* your childhood food; you will live *like* the Scouts you used to be. When W. H. Auden said, 'My face looks like a wedding cake left out in the rain,' we see something new and true about both Auden's face and wedding cakes. That sort of good metaphoric language makes our experiences more spacious, expanding our sensibility. In this way a metaphor is a door that appears and opens in the middle of life, giving us new vistas on the world. But the collapsed metaphor of nostalgic simulation tends towards the opposite: to try to step back inside the past in this way seems likely to teach little besides the ultimate irretrievability of childhood feelings. There is the risk that a forty-year-old dressed for camp and sleeping in Boy Scout cabins will take on a Miss Havisham quality, locked inside the past and sitting beside her bride-cake left out for the spiders.

Most claustrophobic of all in this detox experience is the way in which the digital life from which campers attempt to escape pursues them into the woods. For although its visitors insist that a return to real life can be found in this secluded, retro artifice, Camp Grounded is hounded by contemporary analogy, its promises licked by the very tongue it seeks to dodge. 'Disconnect to Reconnect' is the camp motto, and would-be campers are told that 'the most important status we'll update is our happiness'. Neither saying would have meant much to a 1950s Boy Scout. Once there, why not post a message on the 'Human-Powered Search Engine' – a physical wall of handwritten and typed messages? Like the Russian government in the aftermath of Snowden and WikiLeaks, Camp Grounded is reverting to typewriters. So, despite the sacks of checked cell phones and tablets, digital language has snuck in, just as in another house in the deep woods, a wolf climbed into Granny's nightclothes. In the words it uses to describe itself, Camp Grounded makes use of both reactionary and progressive imagery. Thus, under the canopy of the sequoias, two sets of metaphors collide in anachronism – the nostalgic and the zeitgeist, the banished online world returning for a good old-fashioned panty raid of those Boy Scout cabins.

Getting to the Greenwood

In 1913, after being memorably touched on the bottom in Derbyshire, E. M. Forster began to write his novel *Maurice*. In the book's Terminal Note, written in 1960, he described *Maurice* as belonging 'to an England where it was still possible to get lost'. The militaristic demands of two world wars, the extensive mapping and patrolling of England in the name of security had, for Forster, robbed the island of its wild places. In *Maurice*, the prospect of being lost in 'the greenwood' provides the book's homosexual lovers with their dream of escape from society's punishments. It is Forster's most claustrophobic novel, and effuses a desire to fall from the low-tech but not unoppressive Edwardian grid. These are characters who want out. This instinct for withdrawal and obscurity certainly resonates with present critiques of digitised life. In a world increasingly mapped and patrolled by online analytics and social media's expansive vision, *Maurice*'s central political dilemma still resonates: how might you not be in hiding, while not being on display?

Forster shades the novel with two opposing types of obscurity. In *Maurice*, to borrow from *Brass Eye*, there's good darkness and bad darkness. The book begins with the bad, confining sort: Maurice's adolescence is a descent into 'The Valley of the Shadow of Life', Forster's image for the roiling ignorance of puberty in which Maurice and his schoolmates blunder about, unable to fathom their own feelings. At university Maurice meets his Cambridge sweetheart Clive Durham for the first time, in a college room at the end of an unlit corridor. We're told how 'visitors slid along the wall until they hit the door'. Never getting out of the murk, in *Maurice*, becomes a moral and spiritual failure. After Clive eventually marries

Anne, Forster portrays their marriage as an extended fumble in the dark: 'He never saw her naked, nor she him. They ignored the reproductive and the digestive functions . . . the actual deed of sex seemed to him unimaginative, and best veiled in night.'

A crisis occurs when Maurice realises that, if life drapes a genteel cloth over the budgie's cage, it also offers another sort of obscurity. In this good darkness you aren't drowsy but alert – cat's eyes flashing – and through it you can move unchecked. During one restless night while visiting Clive and Anne, Maurice yearns to become fully invisible to a society that can only ever partially see him. 'Ah for darkness,' he thinks, 'not the darkness of a house which coops up a man among furniture, but the darkness where he can be free!' He is overwrought and half asleep, and so Forster risks giving him purple cravings for 'big spaces where passion clasped peace, spaces no science can reach, but they existed for ever, full of woods some of them, and arched with majestic sky and a friend'. Leaning out of the window he speaks the word 'Come!' into the night. Happily, the night answers, offering Maurice a friend in the form of Clive's young gamekeeper, Alec Scudder, who has been watching from outside. Having found this 'comrade' whose phosphorescence only he can see, Maurice immediately starts planning how they might escape for ever into the shadows. He feels ready for the challenge, believing that 'the forests and the night were on his side'. Scudder is a perfect mate in this regard, for he too 'liked the woods and the fresh air'. When after some misunderstandings and mutual cruelty the two men reunite in London, they wander the streets of Bloomsbury seeking 'darkness and rain'.

For much of the twelve months leading up to his beginning *Maurice*, Forster was in India. While many travel to such places for enlightenment, it was the country's darkness that interested

Forster's novelistic imagination. Towards the end of his sojourn, he became enchanted by caves. The ornate Hindu cave-temples at Ellora were 'the most wonderful thing' he had seen in India, though 'too diabolic to be beautiful'. The chief danger within wasn't the wrath of the godly statues, but the prospect of encountering a leopard. While Ellora fascinated him, he used the plainer set of Buddhist caves in the Barabar Hills, visited a few months earlier, as the site of his 'Marabar Caves' in *A Passage to India*.

Forster began drafting his 'Indian novel' at more or less the same time as he was writing *Maurice*, though he would not complete it for another ten years. The two works-in-progress were clearly nourished by the same imaginative landscape. The liberating, asocial darkness of *Maurice*'s wild woods condenses and intensifies in *A Passage to India* with the 'black holes' of rock in the Marabar Hills, the later novel's obscured heart. In their disorientating, echoing blackness, the Englishwoman Adela Quested believes she has been assaulted by her Indian host. The case is ultimately dismissed, and it's never clear what Adela has experienced. The 'dark caves' repel society's judgements, but even their mystery and otherness are endangered. Adela's scandal occurs because it is easy to be lost inside the system of near-identical caves, and 'in the future they were to be numbered in sequence with white paint'. The mapping of the world continues. In *Maurice*'s postscript, Forster unites the threat to wilderness that stalks both novels: 'There is no forest or fell to escape to today, no cave in which to curl up.'

By 1960, Forster believed that the greenwood, as a space of freedom and anonymity, had long ago 'ended catastrophically and inevitably'. As a result, the world became smaller, more watched, and the very meaning of escape became confined. 'People do still

this knowledge rely on social media to be shared. For Alec and Maurice, the relatively remote boathouse on Clive's estate is their secluded hut, and they instinctively converge there at novel's end. 'So you got the wire,' says Alec sleepily, expectantly, on his friend's arrival at the hidden place. 'What wire?' says Maurice. Although 115,000 people (and counting) 'like' Facebook's cave romance, a tissue of cynicism unrolls in the comments section. The CIA and stalking are recurring themes of the discussion. Less commented upon, if not less felt, is how this version of connectedness hollows out the essence of obscurity, such that we're left with an unhidden hidden place. This 'best thing' about the cave results in a caving-in, with both sense and a sense of escape trapped beneath the rubble.

While eagles crowd the mineshafts and a thousand friends crouch in every cave, a liberating, ambiguous, Forsterian darkness has fallen over certain digital landscapes. In response to a growing consumer concern over the keen vision and sharp memories of networks, two companies, Silent Circle and Geeksphone, have developed the Blackphone, which offers ways to keep in touch and lose yourself all at once. A Blackphone is a smartphone that prioritises the user's privacy, encrypting and fortifying their data, and blocking apps whose often unread terms and conditions may include 'reading' the user's text messages, recording conversations without permission, or revealing their location to third parties. But a more sinister and expansive obscurity resides in the Dark Net. In this online wilderness, user identity is encrypted using a system known as The Onion Router or 'Tor', which itself sounds like a Scandinavian dread deity. Using Tor, a computer in Sydney can masquerade as a smartphone in Prague, and in this way the coordinates of the physical world are scrambled. As in *Maurice*, darkness has no fixed moral tone, since the Dark Net offers a disorientated space for hitmen, hawkers of child pornography and arms dealers, but also political resistance to totalitarian regimes which blanch the regular web with searchlights. Money is exchanged in the Dark Net in the form of anonymous crypto-currencies such as the early star bitcoin, which continually wipe themselves clean of fingerprints. The Dark Net gained some sustained attention in the mainstream media when one of its sites, the billion-dollar drugs marketplace Silk Road, was shut down by the FBI in late 2013. The administrator of Silk Road used the handle Dread Pirate Roberts, nostalgically named after a character in *The Princess Bride*. The man behind the handle was alleged to be Ross William Ulbricht, but a month after he was arrested for hiring hitmen, Dread Pirate Roberts was soon resurrected and active again. Fans of *The Princess Bride* will know that this is true to form, since the point of this pirate is that he is a legend, not an individual, a position in a chain of power rather than a person, and as such doesn't rely on any one particular animator.

The very fact of a Dark Net may be the place where we preserve our sense of wilderness in an ever-networking planet. Reading another pirate tale, *Treasure Island*, as a child, I remember feeling the power, wielded in the opening lines, of withheld whereabouts. The narrator Jim tells us he has been asked 'to write down the whole particulars about Treasure Island . . . keeping nothing back but [its] bearings'. This was also my first exposure to the convention of the severed date: 'I take up my pen in the year of grace 17—'. I can still picture that little slice of nothing cut into the page. There's something both dangerous and peaceful in that dash, a relinquishing of transparency and revelation. Within the first few lines, Robert Louis Stevenson has left a gash in both space and time, revealing nothing but an absence, making room.

If the dash seemed quaint to a boy in a 1980s schoolroom, it seems Jurassic in a Google world of certainties and coordinates.

Social media, too, constantly invites us to fill in these blanks, to populate empty fields with our particulars, our bearings. Every hour you can witness the great joy and solidarity and succour to be had in this form of togetherness, and you might argue that the dangers of obscurity are not worth the liberations. Wild places are by definition unsafe and potentially diabolic. The glimmer in the wilderness may signal the sanctuary of the hermit's hut, but it may also be a pirates' den. At the end of a dark corridor Maurice meets Clive, whose eventual rejection almost tears him to pieces. Maurice ultimately finds in Alec a comrade with whom to share the wilderness, but Forster makes clear that a life as an outlaw takes guts, and that if you want to find a friend you must be willing to face a leopard.

I'm just about acclimatised, now, to lifting up the duvet and watching at least one white, rubbery cable rise with it, to climbing into bed beneath some sort of wiring. It isn't ideal. A new advantage to being single is that, at any one time, there are fewer devices charging on the covers. A couple's evening wind-down now resembles a pair of roadies setting up the stage for the big gig. But the personal constrictions of living with digitised objects are part of a larger syndrome of claustrophobia, a wider lament for free, unbound space.

Ahead of the 2010 Copenhagen summit on carbon emissions, Jeanette Winterson wrote a short story from the perspective of a polar bear sitting at the top of a warming world and imagining its own apocalypse. 'You see, when I lived far away, you knew I was there, and I kept something for you, even though you had never seen a polar bear or an ice floe . . . I kept your wild, cold, raw.' When all of its kind are gone, this polar bear believes, the dream will remain in the heads of the civilised. By then the civilised will have completed their 'Enclosure of the whole world', where the wild places have been bought up and sold off: 'You started to think polar bear thoughts about icyness and wildness. You went shopping and looked at fish. At night you dreamed your skin was fur.' In the same way, the four-dimensional human, coiled in phone chargers and power cords, dreams of a hermitage in an emptied landscape.

When the reality of an unmonitored, free 'yonder' becomes difficult to imagine, with everything corralled for better or worse within the bounds of whatever we might call 'the grid' – the social contract, the social network, globalisation – then the ancient binary of civilisation and wilderness begins to fold in on itself. The loop is closed. With the loss of this key bipolarity, other binaries similarly break down. And so we begin to inhabit a space in which both YES and NO effect the same outcome, and OUT and IN both mean IN. The beginning of a journey becomes the journey's destination, START = FINISH. It reasonably feels, then, as if there is no way out. Hence claustrophobia: a sense of confinement, a loss of space. There is no longer room in such a closed system for both a thing *and* its opposite, so they save space by huddling together in an absurd clinch. In such cramped conditions, language chases its tail, producing its own absurdities: the static wanderer, the small-scale corporate giant, sanctioned disobedience, exceptional normalcy, the advertised hidden place, the Google-mapped Greenwood. In other words, we suffer a loss of imaginative landmass. The wilderness is a place for outlaws, for moral alternatives, and a biodiversity of possibility. Without a deep, absent, spacious wilderness, a space for zeroes, the binary code becomes corrupted and the poles begin to melt.

7

The Blank Screen

In Robert A. Heinlein's 1941 story '— And He Built a Crooked House', the architect Quintus Teal accidentally constructs a four-dimensional home. He takes his clients on a bewildering tour through the house, where space is arranged in 'substantial and impossible' configurations. They think they are walking out of the front door into the garden and emerge into the lounge. The view from a window doesn't show the expected Californian landscape, but a ground-floor room instead, which somehow 'managed to be in two different places at once, on different levels'. Through another window they can see, impossibly, the Chrysler Building and views into Brooklyn. When they hear voices in other, distant rooms, it takes them a while to realise that they're listening to themselves. Teal tries to chase down an intruder, but 'he could not seem to cut down the four-room lead that the interloper had started with'. This intruder drops a hat, and Teal picks it up to find that it's the same as his own and stitched with his own initials.

Shortly after this solipsistic pursuit, an earthquake rocks the house. The new owners and the architect leap through the nearest window, and as they gather themselves they realise that the house has disappeared: 'It was gone. There was no sign of it at all. They stood in the center of a flat desolation.' The sand beneath them seems Martian, but it turns out that they have been transported to Joshua Tree National Park in the Mojave Desert. They hitch-hike back to find that the house has absolutely vanished, knocked by the quake, 'into another section of space'.

The desert likewise appears as a significant landscape in our four-dimensional times. For those who dream of hermits' huts and bite their cheeks in their sleep, the desert represents the ultimate, unmapped wilderness. It exists, in the western imagination at least, as a landscape of erasure. If your aim is to be lost, there is no place more hospitable. In his theorising of America's deserts Jean Baudrillard extrapolates that a desert is 'an extension of [human- ity's] capacity for absence, the ideal schema of humanity's dis- appearance'. It can be a tricky thing for Europeans to talk about deserts in such magnificent generalisation. Too much awe can take you uncomfortably close to a kind of orientalism that gets its kicks from exoticising deserts, particularly those in the Middle East and Africa. In this sort of imagining, these actual, geographical deserts are seen as realms of mystery and magic, and altogether lacking any prosaic, dignified history and ecology of their own. The landscape can almost suffer with weary bemusement this sort of romanticisation, but its people shouldn't have to be swept up in silks and wiles. Yet we all have the right to an abstracted, unpeopled idea of the desert, an empty place, an opportunity for disappearance, stretching for ever in every direction. This abstrac- tion belongs to each of us as much as do our shadows. The idea of a desert wilderness exposes us to the rudiments of ourselves as creatures of space and time, pushing these dimensions to their extremes to produce a boundless, timeless territory of the imagination.

Today, the desert flourishes as both fact and figure of our physical world's narrowing, but at the same time the hyperdrive of digital progress seems to challenge the very possibility of deserthood. Therefore, do we live in the age of deserts or no deserts? The desert, as we have seen, is the outlaw's home, a place beyond the grid. And so, as digitised beings our habitat is the curious landscape of a shrinking expanse, which bleeds over the horizon towards our feet, while retreating in our minds. To put it another way, it is as though we're driving towards something that gets smaller as we approach it.

The Path-Maker King

The rain was heavy when I first heard that Google was expanding its Street View to the desert. I was checking my phone in a dark London square, my fingers keeping the night out of the screen. It was an evening when portents come more easily than buses. Google had on that day revealed its images of the Liwa Desert, an oasis region in the United Arab Emirates, thus ushering this place into digital being. A camel with a camera strapped to its back had gathered the footage.

As in life, on Street View I'm a terrible navigator. Typically, when attempting the anticlimax of looking up my old homes, I make the strangely unaffecting street swim round a few times, as if nostalgia and I are leaning back and spinning on the axis of our clasped, outstretched hands, before something snaps and I swerve face first into a patch of asphalt, as though trying to burrow my way through to the sewers of my youth. With Google's Liwa, if I play it safe and keep clicking just above the horizon with the white magician's glove, then I can make the camel inch across the dunes, each step a blurry lurch. Its fleetness is no doubt curtailed by my broadband speed. Traditionally, blind guides have been prized in the desert, since they are spared the landscape's visual trickery, and in this sense a camel is a fitting symbolic choice, being as spared as a blind person from the fallibilities of human sightedness.

The googled desert is the digital continuation of a very old desire, which often goes by the name of civilisation, to subdue the wild places of the world by laying roads down through them. Indeed this ability to forge routes through wilderness is bound up in the right to power; an archetypal quality of 'kingliness' is the skill of bringing paths to pathless lands. Finding a way is synonymous with leadership, and can be found in the sandy roots of western culture. Early Christians saw John the Baptist, the *vox clamantis in deserto*, as the one who, in the desert to the east of Jerusalem, prepared the path for the final path-maker. Of this holy arrival, Isaiah prophesies that 'The wilderness and the dry land shall be glad; the desert shall rejoice and blossom like the crocus; it shall blossom abundantly and rejoice with joy and singing.' The arrival of the Lord is signalled by a vast irrigation, in which 'waters break forth in the wilderness, and streams in the desert', but most vitally 'a highway shall be there, and it shall be called the Way of Holiness'. We have tended to imagine the desert as lying in wait for a road. The scholar Matthew Bell, in his study *Melancholia:*

The Western Malady, uncovers in desert writing the corollary to this path-making figure: 'the theme of the dejected prince wandering in the desert'. Bell argues that the 'lack of paths in the desert signifies aimlessness, a loss of direction, and poor leadership'. Thus, deserts are ultimately a test of the ruler, whose claim to power depends on cutting a civilising route through wilderness.

So it has come to pass that Google is our era's royal path-maker, the ranger-king for whom we and our wildernesses have been waiting. This shouldn't really surprise us; the auspices have always been clear. Google's self-declared mission 'is to organise the world's information and make it universally acceptable and useful'. Its original purpose was to forge paths in a wilderness of websites. Each search cuts a road through a useless chaos of possibility. Google's CEO, Larry Page, has conceded that this mission statement no longer accounts for their ever-expanding interests, such as automobiling, artificial intelligence and biotechnology. The digitising of the desert is another vector of the company's growth.

The desert is one of our earthly phenomena that are most undone, most compromised, from being googled. From an outsider's viewpoint, the attempt to create a digital trace, to coordinate, orientate and chart the desert, to link it into the world of the known and the searchable, robs the desert of its essential nature. Thus, with the trailblazer's necessary combination of audacity and wonderment, Google's Desert View has symbolically rendered the digital age's campaign to dissolve the possibility of deserthood. The values inherent to our present mode of digitised life are the opposite of desertification. Connectivity, visibility, ever more intricate networks of pathways, web-presence, ceaseless opportunity for communication – these currencies irrigate our sensibilities. Prosperous westerners inhabit an age in which the fundamental qualities of imagined deserthood – remoteness, solitude, pathless-ness, disappearance, inaccessibility and silence – are becoming scarcer.

In a similar vein to the mapping of the desert, there is a cartoonish irony to the fact that companies such as Facebook, Amazon and Google are setting up home in the desert to build their Clouds. These digital acreages populate Oregon's high desert with machinery, their vast plots of hangars housing thousands of computers that hum with information. Keeping these servers from overheating, in the perverse logic of artificial refrigeration, makes heat and requires energy. These mega-corporations appreciate Oregon's cheap electricity, while the dry air assists in cooling. Spaciousness, too, is an obvious advantage of a desert location. Meanwhile, across the dunes in Utah, the Beehive State, the National Security Agency has built a billion-dollar complex in the desert. There it stores the bytes gathered from its searches, filing the comprehensible and working on ways to decode the encrypted.

But the most spectacular example of recent desert path-making is the earnestly proposed mission to Mars in 2024. Funded with private investment, the Mars One project aims to send four earth-lings to our neighbour planet and is currently recruiting volunteers. The organisers will also build Terran training sites in climates that offer 'realistic desert terrain'. Candidates for the Mars One mission apply with the understanding that they will never return home. However, a major aspect to this technological miracle, should it occur, will be that this intrepid quartet will not be entirely lost to the stars. One shortlisted candidate, Hannah Earnshaw, remarked in a *Guardian* interview that 'Obviously it's going to be challenging, leaving Earth and not coming back. I've had support from my friends and family and we can still communicate via the internet.' Mars is already equipped with a satellite in stationary orbit, which exchanges data with those circling earth. As the project's organisers cheerfully put it: 'Fortunately, there would be no limitations to email, texting or "WhatsApping" with the Mars residents.' In other words, a miracle has already occurred: that long-familiar, rusty twinkle in the night sky has been hooked into the network. An otherworldly desert waits now for its path-makers, and even the heavens are within earshot of our tweets.

Neo-Asceticism

Our sense of digital life's encroachment on deserthood is palpable in the ways we talk about antidotes to it, or how we imagine those who attempt to resist its influences. Alternatives to a networked life are coming to seem marginal, and are increasingly being conceived as a sort of neo-asceticism, a return to an old, desert sensibility. Connectivity makes solitude more vivid by its scarcity, and here too the desert appears, since it is embedded in our language of solitariness. Hermits take their name from the Greek *eremia*, which means 'a desert place'. The associated word *eremos* evokes desolation, abandonment and empty wilderness. As discussed, the hermit, as the three-dimensional human incarnate, doesn't figure in digitisation's story of itself. From the same root we get the word eremite, which describes those who abandon the civilised world to seek religious enlightenment in the desert silence. Eremites first appeared as a nameable subculture of Egypt's fourth-century monks, who ventured into the desert regions beyond Alexandria. There they were athletes in suffering, devoted to an asceticism that they believed brought them closer to the ordeals of Jesus.

These early Christians, of whom St Anthony became a prominent example, felt that in the desert quietude they would be more likely to hear the voice of God. The special quality of desert silence became incorporated into the broader religious culture. Of desert lands, the 1859 *Christian Treasury* claims, 'There is no silence so profound anywhere on earth, either by day or night.' The desert's silence is seen in these terms as the outward expression of an ideal state of mind, one that is open to spiritual enlightenment. Baudrillard gives a secular version of this idea when he writes of the desert as an extension of 'the inner silence of the body'. We find today that the intensifying of digital life has produced, as an explicit remedy to such intensity, its own version of quiet contemplation. Consider the edict, which appeared overnight from our culture's decentralised and disembodied authority on well-being, to be *mindful*, to reclaim our own inner silences. As a modern human quality, mindfulness is at once irreproachable and elusive. Indeed it has come to feel something of a duty, all this mindfulness, like recycling. And, just as Google is excited to incorporate desert images into its planetary archive of 'views', we are finding ways of using digital technology to preserve the very sense of desertedness that it elsewhere crowds out. Ascetic virtues are now downloadable in apps such as SelfControl, which allows you to time-limit your access to certain websites throughout the day, typically in the name of productivity. Another app called Headspace, developed by a former Buddhist monk, delivers a meditation regime directly into your smartphone. Snapchat and other 'ephemeral' social-media platforms erase messages and pictures after a certain period, and can be read as part of a larger impulse to recode digital life with an eremitic sense of disappearance, of absence.

This desire for the blank screen can make those who abstain more generally from digital media resemble, from certain angles, one of St Anthony's eremites. You will likely know such people. How solemn and beautiful and remote they can seem, in their desert outposts, quietly observing an extreme way of being. A sense of this neo-asceticism, as a rare alternative to the habits of digital life, is increasingly remarkable. This type of abstinence crosses party lines between the secular and the devout. I witness people chatting to their priests on Facebook (That's very nice of you to like my picture, Father). In 1998, someone I vaguely knew at university dropped out in favour of Buddhism, and instantly turned suitably hushed and laconic. After years of no contact he tracked me down on Facebook. Though he has yet to develop a lucrative app, he seems to be intent on building a digital empire, with his YouTube subscribers alone numbering in the tens of thousands. His YouTube banner is a picture of him meditating in a glade, three tripods with cameras forming a triangle around him.

By contrast, a thoroughly godless friend, my junior by a week, is one such digital ascetic. His remoteness from mainstream behaviour pre-dates Web 2.0 – all knowledge of Gwen Stefani passed his twenties by – but the digital age has certainly crystallised it. His only social-media manifestation is a desolate, silent Facebook page. In his profile picture he appears as an indistinct near-silhouette, like a figure from a Lowry street scene teleported to an empty shoreline. There he is roused once a year by friends for a few many-happy-returns. One afternoon I watched him sit by the window of his bedroom in Stamford Hill, hungover, smoking, waiting for nothing, glancing at passers-by, at groups of Charedi Jews talking outside, and I remember being struck by an atmosphere he produced that was almost like piety.

It isn't uncommon to hear a lamenting sort of admiration voiced in the direction of people who are genuinely immune to the gleaming advances of social media. Such a lament could be heard one evening at the 92nd Street Y, when Sarah Jessica Parker appeared in conversation with Jonathan Tisch, the Chairman of Loews hotels. The topic of social media arises, and an illustrative exchange occurs. Parker admits to being 'slightly terrified of Twitter, every day, every second'. Tisch replies, 'I don't tweet, I don't instagram,' and Parker smiles wistfully. She confesses to being irked that her dealings with these media have changed her existentially and irrevocably. 'It bothers me that I can never go back . . . I can never be the person that *hadn't yet*.' Tisch, of course, is that person. 'I envy you,' Parker says, 'there's a purity to you.'

'No there isn't!'

'No, but there is yet . . . there is yet.'

For all the lightness of this chat, Parker's postlapsarian imagery does articulate a cultural feeling that digital technologies are producing their own Songs of Experience. These technologies now occupy such a central position in our view of people that, if you're someone who appears to abstain from them, this innocence can trump your other claims at worldliness. In 3D life, hotel magnates are rarely mistaken for monks. A 1988 *New York Times* article reported Tisch's marriage to Laura S. Steinberg as the 'joining of two billionaire families'. In the morality of digital asceticism, it is harder for a Google Camel to pass through the eye of a needle, than for a rich man, etc., etc.

Elsewhere, digitisation's effects on our brains have been sceptically presented in terms of a lost asceticism. In *The Shallows*,

Nicholas Carr suggests how a reader's relationship to the codex book, which produces that quiet space of concentration, brought post-Gutenberg readers into the headspace of the holy. In other words, the rise of the widely available text encouraged the intellectual qualities of contemplation and deep attentiveness theretofore associated with religious life. Once again, this incidental, technologically determined monkishness wasn't a signal of virtue. In this case it reflected a cognitive style, a way of thinking shaped by reading habits. As Carr argues, in the era of the codex one didn't need to be saintly to develop a monastic brain. The particulars of the codex's form contour us neurologically, in ways much different than the types of reading that digital technologies encourage: 'whether a person is immersed in a bodice ripper or a Psalter, the synaptic effects are largely the same'. While many are no doubt irritated by Carr's diagnosis of the shallowness of digital reading and comprehension, it's nevertheless significant that he describes this notion of our lost profundity as a kind of secular devoutness.

The Other Oedipus Complex

With deserts throughout the solar system being networked into proximity, and the digitised person set in opposition to the eremite, it seems as though we have deserthood on the retreat. Matters are complicated, however, by the fact that, from a different point of view, deserts are advancing towards us, from multiple directions at once.

I was first exposed to the idea of the merciless desert as a boy, when I went with my sister to see *Return to Oz*. This was an unofficial family tradition. In the early 1940s my old Celtic uncle was taken to watch the original, and he became so distressed that my grandmother had to remove him from the cinema. 'What's she going to do to the wee girl?' he asked, tears flowing. Half a century later I was a similarly tim'rous, cinema-going beastie, sitting in front of the sequel. If time could wrinkle and pull us two boys together, I would be the younger by a year or so, but I had the social pressure of my older sister's worldly, eleven-year-old friend. Perhaps this made the difference to my staying power as the new witch, Mombi, persecuted the new wee girl. And so I kept my head whilst Mombi pulled off her own and chose another from her cabinet of living busts. Moreover, for the younger boy in that later decade, the witch wasn't the real problem. I had my hands and wide eyes full with a more sublime menace, the sort that makes you stop and stare rather than weep in anguish. More arresting than a green-faced psychopath was the Deadly Desert encircling the Land of Oz.

Dorothy's return begins in this arid place. She jumps from rock to rock, avoiding the sand that would turn her likewise if she touched it. Later a pack of punk-ghoulish Wheelers, after chasing the wee girl, goes careening into the desert's edge on all-castered-fours. A splayed body time-lapses into a sandcastle version of itself. A head with one cheek on the sand, its carved sand-mouth gaping and blind sand-eyes staring, cracks in two at the brow. The forehead cleaves; the face falls to pieces. The Oz desert may be far less famous than the ruby slippers, but it recurs throughout the books. In the story *Ozma of Oz*, the eponymous ruler and her entourage cross the desert on a long emerald carpet that rolls itself up in their wake. Dorothy, of course, is a child of the Dustbowl;

she has left one dry land to arrive in another. The word 'gray' pervades the opening passages of the book in which she debuts, muting descriptions of her drought-ravaged Kansas farm and its surrounding landscape. When the Oz books' author L. Frank Baum lay dying, the desert returned to him. Among his final words were: 'Now we can cross the shifting sands.'

The Wheeler's crumbling head means more to me now than it once did. The image sometimes returns when I consciously think about 'the environment', a subject that is otherwise chronically present as the slightest of feelings, tacitly there in the forgotten hours, in smelling the milk, scanning the burst seams of sandwiches for an answer, in seeing, with neither surprise nor boredom, the front door grow closer and closer on the journey home. Saying 'the environment' to yourself never bodes well, as the all-clear will never sound across that set of thoughts. I continue to think of the Wheeler because in the merry old lands of earth, the great deserts are spreading, chewing at their edges. The Gobi Desert in China is growing annually at 'an alarming rate'. The Sahara is 'crossing' the Mediterranean into southern Europe, though the emphasis of this explanation seems to mirror other xenophobic anxieties. The UN has in the past decade declared the related but distinct process of desertification, the loss of the planet's arable drylands from causes such as over-grazing, as being one of 'the greatest environmental challenge of our times'.

People who now and then think of 'the environment' look up to find UN web pages wedged between other tabs of greater and lesser urgency. They remember already knowing the stats when they reread them: 12 million hectares per year lost to drought and desertification, arable land disappearing at '30 to 35 times the historical rate'. They cling to unfazed heroes – local farmers steadily reseeding Ethiopia, the architects of the Great Green Wall of China planting a barrier of trees against the desert and inspiring a similar, multinational effort across the dry shoulders of Africa.

And so we hit a conundrum. The fourth dimension is apparently reducing our deserthood, while deserts are thriving in the other three. Is our civilisation's advancement bringing us towards or away from the desert? *Eremos* grows larger, but there seems less room for eremites. We have encountered variations of this conundrum before. Indeed, western civilisation is in one sense a history of this circular relationship to desertification, which we see manifesting intensely in our own double-sided era of digital fertility and ecological foreboding. The ancient Greeks, drawing up their dense blueprint of our ongoing concerns, understood how societies simultaneously move away from and towards desertification. Consider one of the most influential Greek tragedies. The story of a man who kills his father and marries his mother is what my friend who directs musicals would call 'high concept'.* Freud was most taken with the high concept of Sophocles' tragedy, but as well as inspiring a model for our repressed lusts and rivalries, the play provokes the less garish but equally foundational human idea that civilisation is caught between two types of desert. The first type is that which it seeks to tame and to civilise, the wilderness that must be mapped and paved. The second is those deserts, both actual and metaphoric, which civilisation produces in its wake: fertile land overworked until it is barren earth, but also the figurative desert of the desolate city, the urban wasteland. This transit between deserts is the Other Oedipus Complex, and our expansion into the extra

---

* Another example he gave me: 'School band wiped out in coach crash, come back as ghosts to teach the new band how to win.'

dimension of the digital has certainly intensified its effects on these times.

Here is the priest's account of the suffering city of Thebes, in the opening of *Oedipus the King*: 'A blight is on our harvest in the ear, / A blight upon the grazing flocks and herds'. The royal, damned house of Cadmus is 'emptying'. The priest begs Oedipus to shatter this curse, for surely he would prefer 'To rule a peopled than a desert realm'. Oedipus echoes this image soon after, describing Thebes as 'A desert blasted by the wrath of heaven'. The collapse of civilisation is rendered as a flourishing of barrenness. King Laius, among other things Oedipus' predecessor, has been murdered, and Oedipus realises that the desert curse plagues them because the murderer has not been brought to justice. So he launches that fateful inquiry in which he is both detective and criminal, both the mystery's beginning and its end. Indeed, the blind seer Tiresias tries to explain this circularity to Oedipus: 'I say thou art the murderer of the man / Whose murderer thou pursuest.' Eventually Oedipus realises that the aggressive charioteer whom he killed where three roads meet was also the man who sired him and ordered his abandonment on the 'trackless mountain side'. His uncle/brother-in-law Creon is an astute enough politician to seize on his opponent's incest and regicide, and in any case Oedipus is willing to submit to the punishment that he himself set for Laius' killer. Having gouged out his own eyes, he asks Creon: 'Set me within some vasty desert where / No mortal voice shall greet me any more.'

Sophocles' play has become a textbook illustration of dramatic irony, since antique audiences were familiar with the Oedipal myth and therefore knew, long before Oedipus, the dogged cruelty of destiny. By attempting to dodge fate Oedipus' parents abetted it, an irony that has become notorious. Less recognised, however, is the other strange irony that in curing one desert Oedipus is condemned to another. This is in a sense another expression of the loop in which Oedipus has always been trapped. As Tiresias warns him: 'Thou thyself are thine own bane.' But the next wave of Theban rulers is also exposed to this strange oscillation between civilisation and barren wilderness. King Creon, by favouring edicts over democracy, effectively empties Thebes of its citizens by refusing to honour them as political agents. 'That is no city which belongs to one man', Creon's son Haemon tells him, adding that 'As monarch of a desert thou wouldst shine.'

The criminal Antigone, meanwhile, who in some versions of the myth wanders with Oedipus during his desert exile, is punished for defying Creon and burying her brother. Creon's sentence is to 'take her where the path is loneliest, and hide her, living, in rocky vault, with so much food set forth as piety prescribes'. Deserts are imagined as pathless places, but the loneliest path surely approaches them. And although the image of the desert isn't introduced explicitly until Friedrich Hölderlin's translation, Antigone has always evoked it, covering her brother's body with arid earth. The guard who reports the illegal burial tells Creon how 'the ground was hard and dry, unbroken, without track of wheels; the doer was one who had left no trace'. This trackless, arid crime scene is thus figured in classic desert terms, as part of a landscape where traces disappear as the wind reorganises the dunes. It is only when Antigone returns to her brother's body that the guards catch her.

Who, then, is the desert-dweller? Is it the ruler Creon, whose choking tyranny makes a desert of the agora? Antigone, scatterer of dusty soil, brings the desert's lawlessness to the city, and in doing so is banished to the deserted wilds. This sandstorm blew in with Oedipus, and in one form or another continues to swirl through the ages.

A clear line can be drawn between our version of this complex – the simultaneous sandstorm and Twitter storm of digital late capitalism – and the industrialising nineteenth century, a period particularly agitated by this desert conundrum. With industrialisation and urbanisation came the twinning of progress and desolation, such that the newly founded, teeming metropolis could not be extricated from the image of the wasteland, those vast, human-made expanses of concrete. Nineteenth-century Paris's self-named 'demolition artist', Georges-Eugène Haussmann, struck his wide, civilising roads through the tangled knot of the city's medieval alleys, but in taming one wilderness it seemed he created a barren other. The architect Le Corbusier once described a fight between Haussmann and the Parisian Chamber of Deputies: 'One day, in an excess of terror, they accused him of having created a *desert* in the very center of Paris! That desert was the Boulevard Sébastopol.' At around the same time, Dickens wrote this description of London in *Our Mutual Friend*, which is not atypical for the period's imagining of the city: 'Between Battle Bridge and that part of the Holloway district in which he dwelt, was a tract of suburban Sahara, where tiles and bricks were burnt, bones were boiled, carpets were beat, rubbish was shot, dogs were fought, and dust was heaped.' Here we see London both ingesting and decomposing its own debris, working away at itself, and also forging itself anew, but we aren't allowed to forget the great heaps of dust that are an inevitable outcome of all this industry.

In America's nineteenth-century expansion, Eadweard Muybridge's panoramic photographs of 1870s San Francisco strikingly visualise this Theban curse of modernity. Three decades previously, the Gold Rush had irrigated the region, and San

Francisco became the West Coast's largest city. In the summer of 1877, Muybridge stood on Nob Hill and produced a 360-degree panorama. Once developed it was flattened like a world map, the seam between west and east unstitched and the city unrolled. The pictures show the scale of this boomtown: countless rooftops coloured in the duns and creams of history. Houses, factories, strangely too various for the outsider to distinguish, run as one into an indistinct stipple of sepia at the bay's edge. This downhill slide from ornate Lego brick to textured smear – a dense cascade of wall-roof-wall – is an aspect of modernity's grandeur.

And yet Muybridge's panorama is tactless, revealing the thing that dogs the good times. For between the rows of walls and rooftops are the dusty wedges of streets, and they are empty. The boomtown is a ghost town. Muybridge had hit a technological ceiling. The hours of exposure needed to catch this image meant that his camera's eye slowly dissolved the city's crowds over the course of the day. You may see a small blur of someone here and there, four times their true width and pulled into translucency. In general the roads are swept clean of life. It would be eighteen years until Wilhelm Roentgen took the first X-ray, which was tactless in its own way for unearthing the shadow of his good lady wife's pelvic bone. But up Nob Hill, Muybridge created his own memento mori, revealing the desert beneath the city's skin. His pioneer's lens, absorbing the world at precisely the rate of progress, looked down at the urban modernity below and captured the latent anxiety that our cities are doomed to make deserts of themselves.

Double Vision

We imagine that cities *desert* themselves in two ways. The first is in the more literal, apocalyptic sense, with the fever of industrial urban expansion terminating in a concrete wasteland, its air polluted and greenery long since withered. The second sense is metaphoric, and relates to the homogenising influence of advanced capitalism on city life. Many writers have recorded the popular feeling of a city progressing towards sterile uniformity. This is well illustrated in the phenomenon of chain stores forcing independent businesses out of neighbourhoods by driving up rental prices, so that high streets become echoes of one another, repeating the same narrow selection of franchises. Rebecca Solnit writes about the 'monoculture' produced by San Francisco's gentrification, which many argue is the fault of Google's presence in Silicon Valley inflating the housing market. Fran Lebowitz admires the 1970s street scenes of New York photographer Garry Winogrand in part because 'they were taken before every place became the same place. Before everywhere was nowhere.' It is this quality of uniformity that draws comparisons to desertification. Discussing more or less the same processes as Solnit and Lebowitz, the *New Yorker*'s Adam Gopnik has described New York's vivacity as wilting in a 'monocultural desert of sameness.' Thirty years previously, Richard Cobb wrote nostalgically of Paris that 'The Old Halles is a sad desert.' His melancholy for a lost, lively past drifts and settles across his view of the future: 'There will be no more novelists of the VIe [arrondissement], for there is no life there any more to write about.'

Digital life is in this sense a new kind of boomtown, an opportunity for intense fertility and vivacity, with web pages proliferating in an ever-expanding universe. It wouldn't be controversial to describe mobile communication, alive with tweets and chimes, using the ultimate urban adjective of 'bustling'. There we gather in flocks, faces tumbling over faces. Babies crawl up and down the timelines, while links and memes fill the sidewalks. It is a place that's conscious of population and popularity, and broadcasts tallies of both. There's space for everyone, as follower counts click higher and higher, and almost every new socially mediated offering gains a constituency. It is the rare post that is unpeopled. In the surrounding countryside, the data blooming in the mulched soil isn't soberly collected, in the old bureaucratic style, but harvested.

And yet, even here in the greater digital metropolis, there is unease that all this activity moves us towards unwanted sterility. Zadie Smith argues that social media can enforce uniformity, observing the 'blandness' of selfhood needed for Facebook's vision of openness to be achieved. She writes how, at one stage in the ongoing power play of its privacy settings, 'Gay kids became un-gay, partiers took down their party photos, political firebrands put out their fires. In real life we can be all these people on our own terms, in our own way, with whom we choose.' While Smith feels that in the real world we can be all sorts of people, online it takes much management not to manifest as a uniform singularity. On social media we can build subgroups within groups of friends and followers, allowing information to flow to some and not to others. But generally it has become the standard to upload the single most broadly pleasing version of ourselves, the best shot, which inevitably entails a constricting of personality. We seem collectively to have decided that it's in everyone's interests if we generally present our finest features, cutting a one-dimensional line through the wilderness of the self. Both vanity and expectation are satisfied by this decision.

If we're to accept that people are indeed going online and feeling prone to a groundbreaking sort of sorrow, the kind that makes them dream of the wilderness, then perhaps this sorrow is due to the strain of being a singularity, and from being surrounded by other singularities. It isn't that online we're always happy, or joking, or feeling one thing. 'I'm saying goodbye to my mother today' was one memorable update from an out-of-touch friend. Nevertheless there is an undeniable pressure always to be good, in a singular sort of way. The right kind of righteous anger is encouraged, and how we perform it online is hooked in, as demure as a TV cable, to the backs of our self-esteem. Social media gets a bad rap for incubating egomania, but its patron saint is the superego. We are hyper-conscious of accepted customs and behavioural laws in our little corner of the circuitry, a locality constantly on the cusp of going global. Social media's storms and viruses are effective whips that can discipline our behaviour, limiting the style of our digital personae.

There are therefore parallels between the dreaded monoculture of 3D street life and the present trajectory of our four-dimensional lives. Here we return to some of the ideas discussed throughout the book. Recall the 'chain store' self, representing the pressures on us to develop an online brand that is consistent and virtuous, in order to be part of emerging digital economies. We face demands, both explicit and implied, to gentrify our web-presence, to make it tourist-friendly. In the current paradigm, online life is one of high definition: our updates are written in lights over our shopfronts, each post sharpening the lines of our profiles. We lay down routes through our digital lives for others to follow, the paths well lit and well signposted. On Twitter, hashtags work to standardise our politics, crisply defining and unifying our positions while also promoting a restrictive sameness and uniformity of response. Meanwhile, for those who try to sell us things, our movements through the online world are part of an ongoing process of definition, whereby our tracks through cyberspace become the map of our consumer-selves, the lines on an ever-resolving portrait.

This is how Oedipus has migrated online: in the tension between our boomtown connectivity, our complex, prosperous fertility and the metaphoric desertification of online being that works towards an unproductive uniformity. Such metaphoric use of the desert to evoke a uniform landscape is, however, a common and long-standing failure of the imagination. In 1751, the antiquarian Richard Wood, during travels to Egypt and Syria, wrote un-appealingly of the desert's relationship to creativity: 'a universal sameness of soil, and a constant serenity of sky, afford nothing to awake the fancy or rouse the passions'. By contrast, he felt that Greece was a cradle of culture because of 'varieties of her soil' and the 'vicissitudes of her seasons; and we shall not think it extraordinary, that the arts of life should begin in one of those countries'. Similarly, when a friend called to ask my thoughts on the Desert Street View, she asked why anyone would want to see it, since 'it all looks the same'.

And yet we already know that we simultaneously hold an opposite view of the desert wilderness as a place for alternative ways of being. In this sense, our very idea of the desert is double-sided. E. M. Forster writes metaphorically that 'Great men produce a desert of uniformity around them', while also lamenting the loss of the unmapped desert space, a landscape of freedom and choice, in which his beloved outlaws might stretch out together. In this guise, the desert holds a valuable lesson in the profits of ambiguity, especially to a digital culture that both coaxes and demands us to present ourselves in high definition. While the great servers in America's south-west deserts crunch our data in their warehouses of certainty, it is worth remembering the value of not-knowing.

In Herodotus' *Histories*, from the fifth century BCE, deserts gave shape to the known world's knowledge by being themselves a sort of anti-knowledge. They are the places that resist cartography. In these ancient, western-centric pages, Herodotus verbally tracks the regions of the Middle East and northern Africa – Persia, Phoenicia, Syria, Egypt – giving their relative positions and bounds. Then he declares that 'As far as India, Asia is well inhabited; but from India eastward the whole country is one vast desert, unknown and unexplored.' Likewise, in his account of the geography of Scythia and its neighbours to the north of the Black Sea, he places the Cannibals at the limits of comprehension: 'Above them the country becomes an utter desert; not a single tribe, so far as we know, inhabits it.' Here deserts represent an absence of information. They are impenetrable to scrutiny and cultivate uncertainty: 'so far as we know'. Herodotus' view of a world containing blank spaces in knowledge, which is violent in its own way, is only useful to us now in one sense: it preserves an idea that life can be productively incomplete, making us recall the possibility of unknowability, of inaccessibility. Herodotus' deserts are sanctuaries of privacy, and offer a home to outlaws, those who don't belong to a single tribe.

The Delights of Duplicity

St Anthony sensed how the desert cultivates double vision, a place where demons appear as angels and angels as demons. The desert sensibility, as an unmapped wilderness of the mind, is worth remembering now in situations where high definition obscures rather than clarifies our vision, robbing us of our dimensions rather than increasing them. In a culture drawn into sharp focus by digitisation, the desert is a symbolic place for those who wish to be outliers, if not outlaws. We have always seen the possibilities for alternatives and dissent in such wildernesses. Could *Macbeth* begin in a setting other than where it does, interpreted from the folio directions as an 'open' or a 'desert' place? There the three outlaws gather beneath the thunder, in anything but high definition. This brief scene shimmers with duplicity, not only with the assumed unreliability of witches, but with the more basic sense of duplicity as a doubling. They agree, famously now, to regroup 'When the hurlyburly's done, / When the battle's lost and won.' Immediately Shakespeare establishes the eccentricity of their perspective. The battle will both be lost *and* won; one's instinct may be to take this as a piece of witchy nonsense. But of course that is the thing with battles: they are all both lost and won, and even the winning, particularly, may be equivocal. The history of history shows a preference for the either/or; you lose a battle or you win it. By contrast, the witches have a double vision, which keeps sight of both outcomes at once. The utterance of 'hurlyburly' contributes to this theme as an example of what etymologists call rhyming reduplication. It is a shortening of the Middle English phrase 'hurling and burling', which is really just a playful doubling of 'hurling', an old word for an uproar, a tumult. Their parting chorus exalts in this sensibility: 'Fair is foul, and foul is fair.' This chant could be an echo from the desert ascetics, suffering in a landscape where the difference between angels and demons is unclear. Such ambiguity is the opposite of the claustrophobia of melting poles, in which two opposites subside into identity, where both fair and foul are foul. Here one thing is seen simultaneously as two. When Macbeth and Banquo meet the witches on the heath, their desert place, Macbeth says: 'So foul and fair a day I have not seen.' In other words he occupies the fullness of possibility: standing simultaneously within a foul day and a fair day. There is spaciousness, not confinement, in this eremitic perspective.

In a review for *Esquire Art*, the writer Colm Tóibín mentions his erstwhile doubts about the artist Tracey Emin, having seen her at a couple of parties and thinking that 'she was having too good a time at being Tracey Emin'. Isn't that sense precisely the source of those lonely flares of ill feeling that we sometimes have online, even towards those we like or love perfectly well in the flesh, in the utterly decent longs-and-shorts of them? On social media we see them, as it were, in profile, never looking at us but inviting our gaze, and we believe, in less generous moments, that they are having too good a time being themselves. Tóibín wonders at these parties if Emin's art 'exposed the parts of herself that most needed exposing', and this too is a four-dimensional question if ever there was one. We are put in the position of being the art critics of everyone else's ongoing retrospectives. Why do you think I would want to see that? Who cares about your grandchild? Who cares what your coffee looked like this morning as the sun hit it so dazzlingly? The most common beef we have with each other is that we're not exposing the things that most need exposing.

Imagine Emin's neon artworks as tweets, or status updates. Some of these glowing lines of text would fit in well with a prominent brand of digital spiritualism, embodied in those shared pictures of nature scenes subtitled with uplifting romantic remarks: 'I followed you to the sun' or 'I listen to the ocean and all I hear is you' or 'I can feel your smile'. But then, glowing among these sentiments is the other face of Eros: 'Oh Christ I just wanted you to fuck me and then I became greedy, I wanted you to love me'; 'When I go to sleep I dream of you inside of me'; 'MY CUNT IS WET WITH FEAR'.

You OK hun?

There is remarkable duplicity in this collection. It is hurlyburly in neon, a simultaneity of despair and exaltation, the foul and the fair of Janus-faced love. Tóibín contends that he 'would feel safer' if admissions such as these were ubiquitously flung up in lights, the immanent, urgent and visceral finding shape in a rarefied gas. Referring to Emin's famous tent he writes: 'if more people listed everyone they had slept with, I would somehow rest easier at night'. Reading this sentence in 2009, my first reaction was to think that surely the world is already an infinite campsite of patchwork tents, stitched together with celebrity culture, reality television, social media, confession, exposure. Tóibín must, I thought, be sleeping very well. And yet, Emin's confessional mode exposes how, in the current dynamics of social media, we have sharing aplenty, but far less revelation. I notice that my online colleagues often apologise for 'over-sharing', but as a group we are deceptively single-minded when it comes to the business of confession. We have all the pleasures of confession, with none of the sin. The medium is duplicitous for lacking duplicity. Most of us adhere to a strict style guide of comportment, and for logical reasons. We have absorbed the reality that our digital existence is a book of evidence, and so it behoves us all to look lively.

In this sense, remembering the desert is a solution to our era's claustrophobia. The great triumph of these times will be to preserve, within the pleasures and sorrows of digital life, a collective, desert, double vision. Building paths through wilderness is the first stage of civilisation, but we might ask what sorts of pathmaking are required of us now, in a Google-mapped world? What alternatives to the confining scrutiny of networks, those elaborate systems of roads, can we imagine? In a digital economy where 'trust' comes with a list of ultimatums and a million-dollar insurance policy, we should remember the liberations of duplicity. By keeping the idea of wild, duplicitous places in our minds, they might in turn keep for us the notion of fruitful alternatives, of ways in which we might be cultivated, without being aggressively civilised. In the spirit of this refreshing two-mindedness, we might foster moments when we lose sight of each other, moments when two opposites seem simultaneously possible, a spacious habitat where the unknown parts of ourselves and others are allowed to shimmer in their uncertainty, a humane, unmappable multidimensionality, in the free fall between here and there.

glacier loses parts of itself, splitting into smaller pieces. As a process it is, in the context of current polar feedback loops, the opposite of that which its fecund name suggests. In this idiom birth is a kind of death, procreation amounting to population decline, a moribund baby boom. The metaphor appeared in official English in 1837, borrowed from Danish, and was no doubt less mortifyingly ironic during the wild abandon of the chimney-sweep years. Its English homonym is more intuitive, and more precise. When someone conscientiously takes the knife to the bird, eventually there'll be no more meat on the bone.

The film-maker's description reminds me of *The Inheritors*, of Canterbury twisting and warping and splitting open to reveal an incomprehensible, four-dimensional space behind it. Poetically at least, there's a dark resonance between the apocalyptic cracking-up of the natural world, which itself, unsurprisingly, evokes metaphors of civilisation's undoing, and the ways in which digital life contorts and collapses our spatial certainties, all those old, familiar ideas about solidity. Environmental crisis, to whatever extent we are agents of it, is, after all, the shadow side of our connectivity, a problem of our collective, globalised failures and ignorance.

Christopher Isherwood described the desert as 'a great empty picture frame, and we can't resist using it for a portrait of our private disaster'. Those who looked out on the sublime whiteness of the calving glacier couldn't resist seeing in it a portrait of our collective disaster, the disintegrating city. Likewise, when staring into the illuminated frame of the digital revolution, it can be tempting to inscribe it with some sort of apocalyptic image. A book such as this, which deals with the unfolding present, attracts the Demon of Melodramatic Prophecies. Every book has at least one demon who threatens to make its vision less interesting. I tried to ward off this spirit whenever my thoughts began to resemble a certain run of public-service announcements from the mid-1980s. You may remember them. They starred 'The First Natural-Born Smoker', a tall, pale wretch of a man who lived in a sort of laboratory that seemed to be permanently on fire. There he sat in the fug, smoking, naturally, while a doctorly male voiceover told us about his adaptive physiology. The First Natural-Born Smoker has larger nostrils for inhaling, an elongated index finger for tapping ash, tiny misshapen holes for ears ('Because he doesn't listen!'), extra eyelids to filter the smoke.

While assembling these chapters I would think of the alarmist mood of those adverts. I imagined what a book called *The Four-Dimensional Human* might conclude about our digital evolution. 'The Four-Dimensional Human' will be born with a cupped hand; the long index finger they will share with the Smoker, but soon it will have lost its fingerprints, wiped clean from endless browsing. The Four-Dimensional Human will not keep their memories inside of them; they will have no voice box ('Because they don't talk!').

But how might one reasonably identify a Four-Dimensional Human? The current palm-held system of portable connectivity does coax us into a classic melancholy pose – the solitary figure, eyes downcast, head bowed, a monk in the throes of accidie, the noonday demon. But that doesn't mean we're not having the time of our lives in there. If we were to discern any visible, three-dimensional change, it would be a change in the eyes, the look of pupils that want to dilate and contract all at once, eyes that stay still to see sideways. These are eyes confronted with the considerable and the negligible and for whom the periphery is seen head-on. Eyes that are braced for a camera flash, while at the ready to press against the peephole's tight swirl of light and dark. One may see in these four-dimensional eyes the look of someone who feels a tremor running up their legs, a vibration of the ground beneath their feet, as it shrinks and expands.

I see a look, sometimes, in the eyes of one of my peers – the 1970s and 1980s children, who share my small slice of historical situation in a large, variable world. I think I recognise the homesickness of this look, since these can be homesick times. Everyone has their troubles, but the digital revolution struck this generation at more or less the age when one is expected to put away one's childish things. So as we did what was decent, packing up our most conspic- uous immaturities, the digital revolution began to unravel the very texture of our kind of childhood for all the children still to come. Or so it seemed.

One strand of this texture is entwined for me in a persistent, infantile memory. A children's programme I didn't like was finishing, about hillbillies and soapbox racers, but the credits floating up, up and away were melancholy nonetheless. For it was six o'clock; the child's hours were over. Adults gathered, and standing now on the other side of the living-room door I heard the gongs and triumph of the evening news. I ran up the stairs on all fours, the carpet making my fingertips hum. I knew this off- white sadness of mine for what it was, and had a feeling just then of the homes up and down the land, full of children enduring the same let-down, while not caring about them at all. And maybe it was that day or another just like it when, in the midst of that pause in the fun, in the excitement, I stopped still on the landing and thought 'If there wasn't *life*, what would there be?' And in the dark of my mind I saw my grey-haired father smiling at someone just out of shot. Trying to imagine a world on the other side of my three-dimensional one of banisters and carpets and doorways brought its own unsteadying, circular paradox.

To emerge from such a memory – perhaps in a train carriage speeding up the coast, finding yourself back among the usual phone-stroking, both mellow and urgent, as well as a glowing patchwork of dramas and comedies and cartoons, all of which are happening soundlessly – it seems that the species of tedium that made up your earliest years have vanished for ever. You'll never again stand alone on a wide pavement in the quiet backstreets, not knowing what any of your friends are doing, in that world where people are immutably not there, and programmes finish at the same dreary second, up and down the land. On that train speeding up the coast, cleaved with four-dimensionality, you look around for someone to blame, because you realise that you'll never again be depressed in the ways you once were.

I will have begun to sense the unravelling of my childhood textures in the first few years of the new century, when everyone had a mobile but no one had a smartphone. I knew that my chunky little pay-as-you-go had absorbed other household technologies: watch, alarm clock, camera, calculator. And while I didn't imagine that our phones would soon become our televisions and home computers and maps and mirrors and all the rest, I knew, as we all did in those prologue years, that something was afoot.

This change in the millennial air comes to me now as the memory of an illumination. I have an image of myself lying in the dark, knowing that sleep is close at hand because my mind is nothing but a slow drip of unfinished thoughts. Suddenly I'm aware that the room has altered. Opening my eyes I see that part of the bedroom wall has quietly flushed a bluish-green; a glaucous haze rises from the floor. These visitations are still new to me. The iron tongue, wherever it is, has long struck twelve, the bedroom door is closed and the front door is latched, but there's life still in the day. A drunken friend is drunkenly fond of me, or perhaps is simply glad to have a diversion at the edge of a dance floor, eager to look busy and unconcerned. These are, after all, the dance-floor years. Or perhaps it is young Michael Furey, tapping out a song. Or perhaps it is simply the signal of a fully charged battery, a promise of readiness for the morning to come.

Only because it's so late and the bed is so warm do I lie groggily in the company of this will-o'-the-wisp, before the room dips black again. The colour of the visitation reminds me of something from a book, or rather, after the visitation occurs I will think of it, and be reminded of something else. I'll discover that I'm thinking of Iris Murdoch's *The Unicorn*, of a scene where someone is lost in a swampy wilderness:

> There was, round about him in a great arc, almost encircling him, a fainter line of green light . . . It clung to him, coming into being as he trod, and covering his feet and his footsteps with a weird glow. He hated and feared too the message it gave him as he looked at his luminous trail. He had been walking in a circle. Heaven knew where the right direction lay now. Perhaps after all he had better stand still.

For a moment or so I'll enjoy lying still in this exquisite not-knowing, which in later years will be the main thing I remember. I will of course, shortly thereafter, get out of bed and cut a scrambling path across the carpet to see who is there, shutting one eye against the sudden light.

NOTES

_____

Epigraph

vii 'Postman Pat and his Black and White Cat', written by Bryan Daly,
originally sung by Ken Barrie.

Introduction: The Reverse Peephole

xiii 'an everyday concept . . .', see Linda Dalrymple Henderson's study,
The Fourth Dimension and Non-Euclidean Geometry in Modern Art
(Cambridge: MIT Press, 2013).

xiv 'the old gateway', Ford Madox Ford and Joseph Conrad, *The Inheritors*,
(Liverpool: Liverpool University Press, 2009).

xvi 'to prevent an . . .', 'The Reverse Peephole', *Seinfeld*, dir. Andy
Ackerman, Castle Rock Entertainment, 15 January 1998.

xix Homer, *Odyssey*, trans. A. S. Kline, (www.poetryintranslation.com)

Anatomically Correct

8   'gives me an . . .', Richard Osman interview in the *Metro*, 23 October 2012.

10   'I wish to . . .', Eugene McCabe, 'Music At Annahullion', in *The Granta Book of the Irish Short Story*, ed. Anne Enright (London: Granta, 2010).

13   'even in the . . .', Marcel Proust, *In Search of Lost Time, I: Swann's Way*, trans. C.K. Scott Moncrieff and Terence Kilmartin, revised by D.J. Enright (London: Chatto & Windus,1992).

13   'being out, out . . .', Virginia Woolf, *Mrs Dalloway* (Oxford: Oxford World's Classics, 2008 [1925]); 'What solitary icebergs. . .', *The Voyage Out* (Oxford: Oxford World's Classics, 2009 [1915]); 'For in all . . .', Woolf, *Orlando* (Oxford: Oxford World's Classics, 2008 [1928]).

14   'A generation that . . .', Walter Benjamin, *Illuminations* (London: Pimlico, 1999).

14   'How would it . . .', Jonathan Miller, in an interview with Matthew Stadlen, as part of the BBC's *Five Minutes With* . . . series. http://www.bbc.co.uk/news/20066560.

22   'On the internet . . .', a cartoon caption by Peter Steiner in *The New Yorker*, 5 July 1993.

23   'necessary error', see Judith Butler, *Bodies That Matter* (Abingdon: Routledge, 1993). Butler attributes the term 'necessary error of identity' to Gayatri Chakravorty Spivak.

23   'electronic town halls', see 'Two decades of the web: a utopia no longer' by Evgeny Morozov in *Prospect*, July 2011.

33   'the trust economy', see Brian Chesky on building online brands and verifying Airbnb user identity, in an interview with Derek Thompson in *The Atlantic*, 13 August 2013.

36   'We don't think . . .', see 'Airbnb Now Wants To Check Your Government ID', by Liz Gannes in *AllThingsD*, 30 April 2013.

38   'One turned-up palm', Seamus Heaney, 'St Kevin and the Blackbird', in *The Spirit Level* (London: Faber, 2001).

A Different Kind of Buzz

39 The title of this chapter is taken from a line from Lorde's 'Royals' in the album *Pure Heroine* (Sony/ATV Music Publishing).

43 'the long period . . .' see Juan Antonio Ramirez's *The Beehive Metaphor*, which was extremely helpful on the history of apian architecture and for alerting me to Joseph Beuys's passion for bees (Chicago: University of Chicago Press, 2000).

49 'Here it is . . .' 'I, Borg', *Star Trek: The Next Generation*, Paramount Television, 24 May 1995.

58 From Fasti III. Dionysus, as Bacchus Liber: 'Liber rounded up the swarm, confined it in a hollow tree, / and got the rewards of the honey he found.' In *Ovid's Fasti: Roman Holidays*, ed. and trans. Betty Rose Nagle (Wiley: Hoboken, 1995), p101.

59 'I don't believe . . .', 'Jeu d'échecs avec Marcel Duchamp', dir. Jean-Marie Drot, Antenne 2, 1964.

59 'wanted to become . . .' see "Plight", Joseph Beuys interview with William Furlong in *Art Monthly*, no. 112 (Dec. 1987/Jan. 1988).

Style After Substance

61 Marshall Berman, *All That Is Solid Melts into Air*, (London: Verso, 1983).

67 'Virtual girlfriends' are a thing now. Sad or sweet?', by Daisy Buchanan in the *Telegraph*, 26 February 2014.

69 'living situation', see 'Is the IKEA Aesthetic Comfy or Creepy?', by Lauren Collins in *The New Yorker*, 3 October 2011.

73 'Even though he's . . .', Jennifer Nettles, 'That Girl', Mercury Records, 2013; 'I've been holding . . .' Stevie Wonder, 'That Girl', Tamla/Motown Records, 1981; 'Didn't have to . . .', Justin Timberlake, 'That Girl', RCA Records, 2013.

76 'I am trying . . .', see 'You can't drive a Google car off a cliff. Other than that, they're fantastic', by Elizabeth Renzetti in *The Globe and Mail*, 30 May 2014.

77 'Imagine a *Thelma & Louise* . . .', see 'Google has shown that self-driving cars are inevitable – and the possibilities are endless', by Kevin Maney in the *Independent*, 18 June 2014; 'Suppose *Thelma & Louise* . . .', see 'Less Sexy, Better for Sex', by Delia Ephron in *The New York Times*, 5 July 2014.

78 'The best moments . . .', Alan Bennett, *The History Boys* (London: Faber, 2004).

81 'an untilled field', see 'Patrick Collins and "The Sense of Place"', by Julian Campbell, in *Irish Arts Review*, 4:3, 1987. Campbell is quoting an interview between Collins and Brian Lynch. I was first alerted to Collins's ecological thoughts on the Irish imagination by a conversation between Colm Tóibín and Belinda McKeon.

82 'With absolute unconsciousness . . .', Avram Davidson, 'The Sources of the Nile', in *The Avram Davidson Treasury: A Tribute Collection*, ed. Robert Silverberg (Tor Books: New York, 1999).

83 Anna Wintour's assertion that everyone gets dressed for Bill Cunningham appears in *Bill Cunningham New York*, directed by Richard Press, First Thought Films, 2011.

86 Daniel Spagnoli's comments on Normcore quoted in 'This browser extension lets you block everything #Normcore', by Audra Schroeder in *The Daily Dot*, 7 March 2014; Anna Byrne's article on Normcore was published 30 May 2014 on news.com.au.

87 'is the deepest . . .', Walter Benjamin, 'On Language and Such and on the Language of Man', in *Walter Benjamin: Selected Writings, Volume 1*, ed. Marcus Bullock and Michael W. Jennings (Cambridge: Harvard University Press, 1996). English translation copyright © 1978 by Harcourt Brace Jovanovich, Inc.

88 'it's a reaction . . .', see 'LOL-core – the big trend for autumn 2014', by Morwenna Ferrier in the *Guardian*, 28 August 2014.

91   'my normal is . . .', Seth Farbman, 'Gap's Fall Campaign Celebrates Individuality . . .' a press release by Gap Inc., 19 August 2014.

Keeping and Killing Time

94   'Now *all* the . . .', Elizabeth Bowen's letter to Woolf quoted in *The Mulberry Tree*, ed. Hermione Lee (London: Virago Press, 1986).

94   'true conquerors' time . . .', Jean-Paul Sartre, *Iron in the Soul*, trans. Gerard Hopkins (London: Hamish Hamilton, 1950).

96   'The cool thing . . .', from 'Me at the zoo', YouTube video uploaded by jawed, 23 April 2005.

99   'Keep your lions . . .', *The Box of Delights*, directed by Renny Rye, BBC/Lella Productions, 1984.

99   'Now I'm here . . .', *Shirley Valentine*, directed by Lewis Gilbert, Paramount Home Entertainment, 1989.

101 'in that moment . . .', Proust, *Swann's Way*.

102 'as though "1995" . . .', Philip Roth, *American Pastoral* (London: Vintage, 1998).

103 'Cash unfortunately forgets . . .', Jaron Lanier, *Who Owns The Future?* (London: Allen Lane, 2013).

105 The section on Galliano and Gaddafi is a revised version of an essay first published in *Five Dials, vol. 18b: A Spring Postcard*, 19 April 2011. I am grateful to my editor Craig Taylor for his advice on the original article.

106 Charlie Rose's interview with John Galliano was first broadcast on *Charlie Rose*, PBS, 12 June 2013.

107 'a scene of . . .', Isobel Armstrong, *The Radical Aesthetic* (Oxford: Blackwell, 2000).

110 'used a similar . . .', see 'Revealed: How Jimmy Savile abused up to 1,000 victims on BBC premises', by Daniel Boffey in the *Guardian*, 18 January 2014.

113 'the fourth teenager . . .', see 'Man charged over takeaway shooting' in the *Telegraph*, 20 April 2010.

115 'you are under . . .', see 'Only love and then oblivion...', by Ian McEwan in the *Guardian*, 15 September 2001.

118 'the nation's most . . .', see 'Grief and the Cameras', by Andrew O'Hagan in the *London Review of Books*, 3 December 2009.

118 'Flood everywhere . . .' Hayden White, *The Content of the Form: Narrative Discourse and Historical Representation*, (Baltimore: Johns Hopkins University Press, 1987).

120 'Either they love . . .', see 'One trial too many as Katie Price quits jungle', by Rosamund Hutt in the *Independent*, 23 November 2009.

123 'in great strides . . .', H.G. Wells, *The Time Machine* (London: Penguin Classics, 2005 [1895]).

Weeping Toms

132 'not knowing who . . .', Elizabeth Bowen, *The Heat of the Day* (London: Vintage, 1998 [1948]).

134 'Facebook is slowly . . .', see 'Facebook "Headed For Social Media Graveyard"', by Grace Beadle, *news.sky.com*, 23 January 2014.

135 'may have been . . .', see 'The troll in the president's office', by Mike Wendling for *BBC Trending*, 12 March 2014.

135 Part of this chapter's introduction to Gothic literature has been reworked from my *Free Thinking Essay* for BBC Radio 3, first broadcast 10 November 2011. A thank you to my producer Eliane Glaser for her attentive work on the original script.

143 'the guy in . . .', *Scream*, directed by Wes Craven, screenplay by Kevin Williamson, Dimension Films and Woods Entertainment, 1996.

152 'When people think . . .', *Fight Club*, directed by David Fincher, screenplay by Jim Uhls, based on the novel by Chuck Palahniuk, 20th-Century Fox, 1999.

155 'composed of the . . .', William O'Daniel, *Ins and Outs of London*

(Philadelphia: S.C. Lamb, 1859). Accessed via Google Books digital edition.

159 'Kim Kardashian Overtakes Justin Bieber on Instagram Following Fake Follower Purge', by Salvador Rodriguez in *International Business Times*, 18 December 2014.

The Cabin in the Woods

175 You can find out how to meet a Barclays Digital Eagle at: http:// www.barclays.mobi/MeetourBarclaysDigitalEagles/MP1242672689552

176 I owe thanks to Rob Lederer and his work on archives and databases in contemporary American culture for the tip on Catwoman's 'Clean Slate' dreams.

178 'The movies that . . .', *Looper*, written and directed by Rian Johnson, Sony Pictures, 2012.

178 'It doesn't even . . .', *The Cabin in the Woods*, directed by Drew Goddard, screenplay by Drew Goddard and Joss Whedon, Lionsgate, 2012.

180 'If you have . . .', I have used Celia Sgroi's translation of Robert Schumann's *Eichendorff Liederkreis* song cycle.

182 'Have you ever . . .', Buycott's description of its app dates from 2013, and was accessed at www.buycott.com.

183 'The Hand people . . .', *30 Rock*, 'Brooklyn Without Limits', directed by Michael Engler, written by Ron Weiner and Tina Fey, Universal Media Studios, 11 November 2010.

185 'Years before Occupy . . .', see 'Resistance Is Surrender', by Slavoj Žižek in the *London Review of Books*, 15 November 2007.

188 'When we are . . .', Gaston Bachelard, *The Poetics of Space*, trans. Maria Jolas (Boston: Beacon Press, 1994 [1958]).

189 'Although today I . . .', Rumpelstiltskin's smug song is taken from the wonderfully sinister Ladybird edition of the fairytale, retold by Vera Southgate (Loughborough: Ladybird Books Ltd, 1968).

191 Information on Digital Detox and Camp Grounded can be found at

www.digitaldetox.org. This section's discussion is based on the version of Digital Detox's website as it appeared in the summer of 2014. As the book goes to print, the website exhibits some telling differences and an added urgency: The 'era of burnout . . .' is not 'coming to a close', but rather, 'It's *time that* the era of burnout, FOMO . . . comes to a close' (my emphasis). Camp Grounded remains a place 'Where grown-ups go to unplug, get away and be kids again.'

192 'My face looks . . .', quoted in Humphrey Carpenter, *W. H. Auden: A Biography* (London: Faber, 2014 [1981]).

194 The E.M. Forster sections in this chapter borrow heavily from my article that was originally published in the *Guardian*: 'Ode to the Greenwood', 6 July 2013. My thanks to Paul Laity for commissioning and editing this essay. See also: *Selected Letters of E.M. Forster, Volume 1, 1879-1920*, eds. Mary Lago and P.N. Furbank (London: Collins, 1983); *Maurice* (London: Penguin, 2005 [1971]); *A Passage to India* (London: Penguin, 2005 [1924]).

199 On the resurrection of Dread Pirate Roberts, see 'Silk Road Lives', by Denver Nicks in *TIME*, 6 November 2013: http://nation.time.com/2013/11/06/silk-road-lives/ [Accessed: 31 March 2015]; also, 'Ross Ulbricht: Silk Road creator convicted on drugs charges': http://www.bbc.co.uk/news/world-us-canada-31134938 [Accessed: 31 March 2015].

200 'You see, when . . .', Jeanette Winterson, 'Polar Bear'. Published in the *Guardian*'s 2009 Climate Change Special. Now available at www.jeanettewinterson.com.

The Blank Screen

202 'managed to be . . .', Robert A. Heinlein, '—And He Built A Crooked House', in *All You Zombies – Five Classic Stories by Robert A. Heinlein*, (Kindle Edition, 2013). Copyright assigned to the Robert A. & Virginia Heinlein Prize Trust, 2003.

202 'an extension of . . .', Jean Baudrillard, *America* (London: Verso, 2010 [1988]).

207 'Obviously it's going . . .', see 'Mars One shortlist: five Britons among 100 would-be astronauts', by Ian Sample in the *Guardian*, 17 February 2015.

211 'slightly terrified of . . .', from 'Sarah Jessica Parker with Jonathan Tisch', YouTube video posted by 92Y Plus, 10 July 2014.

211 'Candlelight Wedding Joins 2 Billionaire Families', by Georgia Dullea in *The New York Times*, 19 April, 1988.

212 'whether a person . . .', Nicholas Carr, *The Shallows* (London: Atlantic, 2010).

214 'Now we can . . .', quoted in Frank Joslyn Baum and Russell P. MacFall *To Please a Child: A Biography of L. Frank Baum, Royal Historian of Oz* (Chicago: Reilly & Lee Co., 1961).

214 'an alarming rate', 'Fighting Desertification', by Terry Waghorn, www.forbes.com, 7 March 2011.

214 'Sahara is "crossing" . . .' see 'Sahara jumps Mediterranean into Europe', by Paul Brown in the *Guardian*, 20 December 2000.

214 'the greatest environmental . . .' see 'Desertification', www.un.org.

216 'A blight is . . .', *Oedipus the King*, trans. F. Storr, accessed at: classics.mit.edu.

217 'That is no . . .', *Antigone* trans. F. Storr (Cambridge: Harvard University Press, 1977).

217 Véronique M. Fóti's book *Epochal Discordance: Hölderlin's Philosophy of Tragedy* informed some of my discussion of Antigone's deserthood (Albany: SUNY, 2006).

218 'One day, in . . .', quoted in Walter Benjamin *The Arcades Project*, trans. Howard Eiland and Kevin McLaughlin (Cambridge: Harvard University Press, 2002).

220 'they were taken . . .', from 'A conversation with Fran Lebowitz on Winogrand's New York', in *The Man in the Crowd: The Uneasy Streets of Garry Winogrand*, ed. Jeffrey Fraenkel (San Francisco: Fraenkel Gallery, 1999).

220 'monocultural desert of . . .', quoted in *The Cambridge Companion to the Literature of New York*, ed. Cyrus R.K. Patell and Bryan Waterman (Cambridge: Cambridge University Press, 2010).

220 'The Old Halles . . .', Richard Cobb, *The Streets of Paris* (London: Duckworth, 1980).

221 'Gay kids became . . .' see 'Generation Why?', by Zadie Smith in *The New York Review of Books*, 25 November 2010.

223 'produce a desert . . .', E.M. Forster, 'What I Believe', in *Two Cheers for Democracy* (Harmondsworth: Penguin, 1976).

226 'she was having . . .', Colm Tóibín, 'Tracey Emin', Esquire Art 2007–2010.

Epilogue: Fairy Fire

229 'Imagine Manhattan, and . . .', *Chasing Ice*, directed by Jeff Orlowski, 2012.

230 'a great empty . . .', Christopher Isherwood, *Christopher Isherwood Diaries, Volume One: 1939 – 1960*, ed. Katherine Bucknell (London: Vintage, 2011).

234 'There was, round . . .', Iris Murdoch, *The Unicorn* (London: Vintage, 2001 [1963]).

ACKNOWLEDGEMENTS

I am grateful to the Jerwood Foundation and the Royal Society of Literature for their non-fiction prize, which helped make the space in which this book was completed.

Immense thanks are due to my agent Tracy Bohan, for guidance, advice and friendship, as well as to her colleagues at the Wylie Agency. To the desert vision of Tom Avery at William Heinemann, who unlocked this project and who edited the book brilliantly and tirelessly, in the old style. I would also like to thank the entire Heinemann team for their goodwill and enthusiasm for the fourth dimension. I was very lucky to benefit from David Milner's copy-editing talents.

My friends and colleagues at Arcadia University, past and present, have offered both encouragement and much practical support. Thank you to my students, who put up with being quizzed about youthful online habits.

This work has been informed and improved by many conversations, but there are a few short lines in the book whose phrasings I owe to certain people: Jonathan Allen, Jane Darcy, 'Digital' Dan Franklin, Paul Fleckney, Vanessa Scott. Thank you to Eli Davies and Dan Hancox for catching mistakes on the hoof.

The British Library and Senate House Library in London provided data, desk space, and ambience, and should not be taken for granted.

Thank you to Melissa Clarke and Tom Lederer for their cabin by the lake, in which a chapter of this book was written.

Several people have emboldened my sentences at key stages, their voices coming and going in the writing room: Mary Jane Kearns-Padgett, Jean Wilson, Joe Adamson, Jake Smith-Bosanquet, Patrick Walsh, Walter Donohue.

Multi-dimensional thanks to my dear friends in Canada and the UK; to my siblings Anya and Jerome, for the years. And to Rob Lederer, my bravest early reader, for making rooms light up . . . & other stories.